157

Structure and Bonding

For further volumes:
http://www.springer.com/series/430

Aims and Scope

The series *Structure and Bonding* publishes critical reviews on topics of research concerned with chemical structure and bonding. The scope of the series spans the entire Periodic Table and addresses structure and bonding issues associated with all of the elements. It also focuses attention on new and developing areas of modern structural and theoretical chemistry such as nanostructures, molecular electronics, designed molecular solids, surfaces, metal clusters and supramolecular structures. Physical and spectroscopic techniques used to determine, examine and model structures fall within the purview of *Structure and Bonding* to the extent that the focus is on the scientific results obtained and not on specialist information concerning the techniques themselves. Issues associated with the development of bonding models and generalizations that illuminate the reactivity pathways and rates of chemical processes are also relevant

The individual volumes in the series are thematic. The goal of each volume is to give the reader, whether at a university or in industry, a comprehensive overview of an area where new insights are emerging that are of interest to a larger scientific audience. Thus each review within the volume critically surveys one aspect of that topic and places it within the context of the volume as a whole. The most significant developments of the last 5 to 10 years should be presented using selected examples to illustrate the principles discussed. A description of the physical basis of the experimental techniques that have been used to provide the primary data may also be appropriate, if it has not been covered in detail elsewhere. The coverage need not be exhaustive in data, but should rather be conceptual, concentrating on the new principles being developed that will allow the reader, who is not a specialist in the area covered, to understand the data presented. Discussion of possible future research directions in the area is welcomed.

Review articles for the individual volumes are invited by the volume editors.

In references *Structure and Bonding* is abbreviated *Struct Bond* and is cited as a journal.

Banglin Chen • Guodong Qian
Editors

Metal-Organic Frameworks for Photonics Applications

With contributions by

B. Chen • X.-M. Chen • Y. Cui • S. Du • P. Falcaro •
S. Furukawa • K. Hirai • S. Kitagawa • W. Lin •
R. Medishetty • G. Qian • J.J. Vittal • H. Zhang •
J.-P. Zhang • T. Zhang

Springer

Editors
Banglin Chen
Department of Chemistry
University of Texas at San Antonio
San Antonio
Texas
USA

Guodong Qian
Dpt. of Materials Science & Engineering
Zhejiang University
Hangzhou
People's Republic of China

ISSN 0081-5993 ISSN 1616-8550 (electronic)
ISBN 978-3-642-44966-6 ISBN 978-3-642-44967-3 (eBook)
DOI 10.1007/978-3-642-44967-3
Springer Heidelberg New York Dordrecht London

Library of Congress Control Number: 2014931514

Printed on acid-free paper

Springer is part of Springer Science+Business Media (www.springer.com)

Preface

Photonics, the science of light emission, transmission, modulation, manipulation, and detection, are playing an essential role in many fields, including information processing, communication technologies, military, and biomedicine. Compared to electronics, photonics enable the transport and processing of information at the speed of light. The present boom in optical fiber technology is an excellent example of how much impact photonics has on the advancement of our society. Up to now, various inorganic and organic materials have been widely developed for photonics applications, including nonlinear optical materials, luminescent materials, laser materials, photoactive materials, photoconductors, and so on. The research on the photonics materials combines the applied research of chemistry, solid state physics, materials science, and optical and computer engineering and will be the most dominating forces in various fields of science and engineering.

Currently, metal-organic frameworks (MOFs) are being intensively studied as a novel class of hybrid inorganic–organic material with ultrahigh porosity, enormous internal surface areas, together with the extraordinary tailorability of structure, dimension, size, and shape. Although the photonics application of MOF materials is still at the early stage compared with the application for gas storage, separation, and heterogeneous catalysis, the currently available results have unambiguously demonstrated that the design and construction of MOFs for photonics functionality is very active. Because of the inherent advantages of both organic links (easily modify, flexibility, versatility, etc.) and inorganic metal ions (unique electronic and optical nature), MOFs will open a land of promising applications in photonics fields where conventional inorganic or organic materials might not be suitable. The MOF approach can also offer a variety of other attractive characteristics such as the straightforward syntheses, predictable structures and porosities, and collaborative properties to develop new photonics materials and important applications.

Due to the importance as well as the rapid progress of the MOF materials in the field of photonics, it is quite appropriate to publish this first book on MOFs for photonics application. The book presents the most up-to-date source of information on photonics applications of MOFs and not covered yet by other books. In this book, the database of photonics MOFs, the design principles, the unique characteristics of MOFs, and the potential photonics applications are discussed in detail. The results obtained in the last years by the main leading MOF researchers all over the world are presented in this book, thus providing a valid and precious overview on the last developments and moreover on the future innovative applications of the MOF materials in the field of photonics for scientists working in this area.

The book is organized in six chapters and opens with a chapter titled "Design and Construction of Metal-Organic Frameworks" by Chen and Zhang, which reviews the fundamental characteristics, molecular design, and synthetic strategies of MOFs. The chapter titled "Luminescent Properties and Applications of Metal-Organic Frameworks" by Qian et al. elaborates on the luminescent behavior of MOFs and describes the applications of luminescent MOFs in the fields of lighting-emitting, chemical sensors, and medicine. In chapter titled "Metal-Organic Frameworks for Photocatalysis" by Lin discusses the photocatalysis application of MOF materials. Vital et al. then describe the photochemical transformation within MOFs in chapter titled "Photochemical Transformation within Metal-Organic Frameworks". The chapter titled "Metal-Organic Frameworks for Nonlinear Optics" by Du et al. provides a comprehensive review of MOFs that display nonlinear optical properties. Lastly, Furukawa and coworkers in chapter titled "Host-Guest Metal-Organic Frameworks for Photonics" describe the luminescent properties of host-guest MOFs and their potential applications as sensing materials.

San Antonio, TX, USA — Banglin Chen
Hangzhou, People's Republic of China — Guodong Qian

Contents

Struct Bond (2014) 157: 1–26
DOI: 10.1007/430_2013_100

Published online: 4 May 2013

Metal–Organic Frameworks: From Design to Materials

Jie-Peng Zhang and Xiao-Ming Chen

Abstract Metal–organic frameworks are a new class of advanced porous materials, which have several important characteristics, such as highly periodical, diverse and designable structures, high porosity, unique and modifiable pore surface, as well as framework flexibility. In this chapter, we give a brief account focusing on the molecular design, synthetic strategies and interesting properties and applications of metal–organic frameworks with a number of selected examples.

Keywords Application · Metal–organic framework · Molecular design · Porous coordination polymer · Property · Synthesis

Contents

J.-P. Zhang (✉) and X.-M. Chen
MOE Key Laboratory of Bioinorganic and Synthetic Chemistry, School of Chemistry and Chemical Engineering, Sun Yat-Sen University, Guangzhou 510275, China
e-mail: zhangjp7@mail.sysu.edu.cn; cxm@mail.sysu.edu.cn

1 Introduction

Although the term coordination polymer first appeared in 1960s, and can be traced up to the first man-made coordination polymer or Prussian Blue by an accident in early eighteenth century, the very extensive interest in coordination polymers was triggered by several pioneering reports on discoveries of the porous structures and related functionalities around 1990. In particular, Robson et al. reported the porous metal cyanides with ion-exchange properties [1, 2], Kitagawa et al. demonstrated methane adsorption in a microporous coordination polymer [3], and Yaghi et al. reported a rich family of highly porous and thermally stable metal carboxylates [4] that were termed metal–organic frameworks (MOFs) [5]. The research in coordination polymers has grown substantially in the past decade, as being evidenced by the exponential increase of publication number.

MOFs are constructed by metal ions/clusters and organic bridging molecules. They can be regarded as a subclass of coordination polymers, since the metal ions/clusters should be solely bridged by organic molecules (or ligands) via coordination into two-dimensional (2D) or three-dimensional (3D) porous networks in MOFs [5], whereas the bridging ligands may be either inorganic or organic in coordination polymers [6]. Nevertheless, the term MOF has been very popularly used for different types of coordination compounds with 3D to 1D, and even 0D, both porous and nonporous structures. Anyway, we refer MOF to porous coordination polymer in this chapter.

MOFs may have both advantages of inorganic and organic materials [7, 8], but their physical and chemical stabilities are relatively lower than those of carbon materials and inorganic zeolites solely composed of carbon–carbon and metal-oxide bonds, respectively. As being well known that metal ions (or clusters) and organic bridging ligands are diversified in coordination behavior and geometry, the combinations of metal ions (or clusters) and bridging organic ligands through coordination interactions are almost infinite. Consequently, a huge number of MOFs with different functionalities, such as gas adsorption, separation, sensing, catalysis, and luminescence, have been documented. MOFs have some important characteristics owing to their unique compositions and structures, such as highly periodical structures, high porosity, special and modifiable pore surface, as well as framework flexibility. The 3D periodicity or crystallinity benefits the determination of their structures, in particular by single-crystal X-ray diffraction, which is fundamental for a well understanding of the structure–property relationship and better design of new materials. High porosity can be rationally achieved by using suitable framework structure and/or long organic ligands, which is important for adsorption and storage of molecules, catalysis, etc. The pore surface of MOF is decorated by both inorganic moieties and various functional groups of organic ligands, which can be readily tailored to meet a specific property or function, such as selective adsorption, separation, and sensing. Compared to the traditional zeolites and other inorganic porous materials, MOFs are very unique for their highly flexible host framework structures, which are very useful for smart materials with special sorption, storage, separation, and sensing properties [9].

Therefore, as a new class of advanced porous materials, MOFs have become an intense subject of scientific research. As many recent reviews on this field have been published, we will give a brief account focusing on the molecular design, synthetic strategies, and representative properties of MOFs by a series of selected examples.

2 Design and Topology

The bonding strength and directionality of coordination bonds are in between covalent bonds and supramolecular interactions such as hydrogen bonds, which imply that structural prediction of MOFs should be easier than that of a molecular organic crystal. Although it is still difficult to emphasize the term design, there are several strategies widely used to design the framework structures of MOFs.

In these strategies, the net-based approach or reticular chemistry is the most popular and effective [10, 11]. As crystalline frameworks, the connectivities in inorganic compounds, zeolites, and MOFs can be abstracted into periodical nets, or topologies, constructed by nodes and linkers. Since the pioneering works of Wells, who introduced the mathematical description for the structures of a variety of inorganic materials and hypothetical networks [12], a large number of topologies have been known/predicted to date, which can be achieved from open access databases such as the Reticular Chemistry Structure Resource (RCSR) [13, 14]. The mathematical description of a topology represents the inherent connectivity but is usually quite complicated and difficult to index. For example, the net of carbon atoms, connected by carbon–carbon covalent bonds in graphite and diamond have the point symbols 6^3 and 6^6, and vertex symbols 6·6·6 and $6_2 \cdot 6_2 \cdot 6_2 \cdot 6_2 \cdot 6_2 \cdot 6_2$, respectively. Therefore, topology symbols usually borrow names from natural structures/compounds, such as honeycomb (for graphite) and diamond. The RCSR symbols consisting of three bold letters, such as **hcb** and **dia** (Fig. 1), were recommended as the uniform nomenclature [15], which cover the IUPAC framework-type codes for all unique and confirmed zeolites (three capital letters, such as SOD for sodalite) [16].

Network topologies have been used to not only simplify and describe but also design the metal-ligand connectivities of MOFs. One can select metal ions/clusters and/or organic ligands to mimic the node geometries of selected topologies (Fig. 2). However, a given node geometry can be always suitable for construction of at least several different topologies [17]. Therefore, simple and high-symmetry network topologies are regarded as the most viable prototype of MOFs, because they reduce the number of possible products of self-assembly. For example, combination of two-coordinated coinage metal ions such as Cu(I) and Ag(I) with imidazolate derivatives give 1D zigzag chains, helixes, or 0D rings. On the other hand, if 1,2,4-triazolates were used, coinage metal ions would tend to adopt trigonal-planar coordination geometries, giving 3-connected topologies. The highly controllable coordination mode and self-assembly behavior of azolate anions (such as imidazolate, pyrazolate, and triazolate) make metal azolate frameworks (MAFs) a unique subset of coordination polymers [18]. While more common ditopic neutral bridging ligands such as dinitriles and diamines were used, coinage metal ions often showed tetrahedral

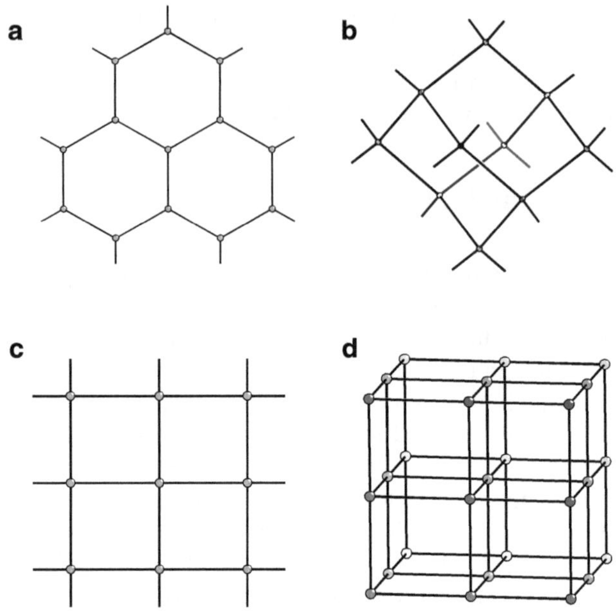

Fig. 1 Selected simple topologies. (**a**) **hcb**, (**b**) **dia**, (**c**) **sql**, (**d**) **pcu**

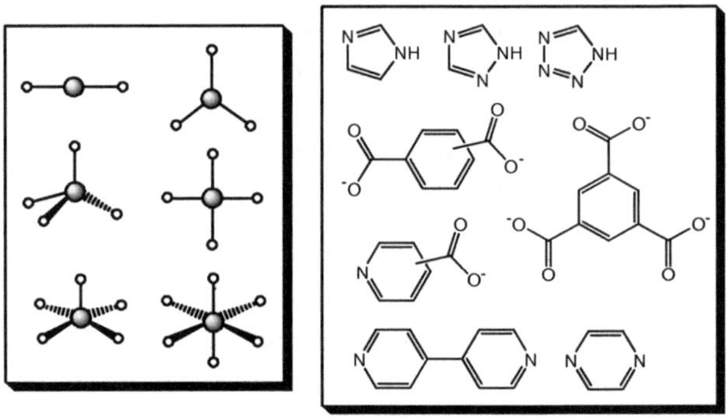

Fig. 2 Typical coordination geometries of metal ions and some simple bridging organic ligands

coordination, producing 4-connected **dia** networks. For divalent transition metal ions, square-planar, tetrahedral, and octahedral coordination geometries are quite common, which are suitable for the construction of **sql**, **dia**, and **pcu** topologies. The 2D **sql** topology can be rationally constructed by rigid linear bipyridyl-type ligands and Ni(II) or Cu(II) ions, in which the axial positions of the octahedral metal ions are occupied by solvent molecules or small counter anions [19, 20]. On the other hand, **pcu** networks based on single metal ions are relatively few because six pyridyl rings can induce significant steric hindrance effect [21]. The **dia** topology may be the most frequently encountered structures in MOFs because of the abundance of tetrahedral metal ions and linear ditopic bridging ligands [22]. Although some interesting porous compounds are based on the common **sql** and **dia** topologies, low-dimensionality and/or interpenetration usually limit their porosities.

Based on the same square-planar node, the 3D **nbo** and **lvt** topologies are more porous than the 2D **sql** one. However, mixing square-planar metal ions with linear bridging ligands generally produce the 2D **sql** network rather than the two 3D counterparts, because low-porosity structures are more energetically stable. Actually, MOFs with the **nbo** and **lvt** topologies are very rare. Comparison of the three typical topologies show that their difference relies on the dihedral angle between adjacent square-planar nodes, in which they are coplanar, perpendicular, and with an acute angle in **sql**, **nbo**, and **lvt**, respectively. To rationally construct the 3D 4-connected networks based on square-planar nodes, control over the node-to-node orientation is necessary. For example, Yaghi et al. rationally designed a highly porous MOF using square-planar $Cu_2(RCOO)_4$ clusters and a bromine-substituted benzene-dicarboxylate ligand, in which one of the carboxylate group is perpendicular to the benzene ring because of the steric hindrance of the bromine atom [23]. With shorter bridging ligands, one may gain more control over the network topologies. For example, we have synthesized a series of Cu(I) 1,2,4-triazolates showing the **sql**, **lvt**, and **nbo** topologies [24, 25]. In the first glance, these coordination networks can be simplified as 3-connected topologies because the metal ions and ligands are all 3-coordinated. But as reflected by their RCSR symbols, **sql-a**, **lvt-a**, and **nbo-a**, they can be also described as 4-connected topologies based on square-planar binuclear $Cu_2(Rtz)_2$ (tz denotes the triazolate ring) nodes. Actually, **sql-a**, **lvt-a**, and **nbo-a** are obtained by 1,2,4-triazolate derivatives with different substituent groups, in which the unsubstituted ligand give the nonporous **sql-a** layer, the methyl substituted one give nonporous 2-fold interpenetrating **lvt-a** network, and larger substituents including ethyl and propyl groups give porous non-interpenetrated **nbo** frameworks. In their crystal structures, steric hindrance between pendant (non-coordinated) groups can be observed. It should be noted that the relatively porous compound Cu(I) 3,5-diethyl-1,2,4-triazolate (MAF-2) has demonstrated a number of interesting sorption properties by virtue of its unusual flexibility on the coordination network and the side groups [26].

2.1 Cluster-Based Frameworks

Obviously, the easy rotation of molecular building blocks around individual coordination bonds is not beneficial for control over node-to-node orientation, which is very important for construction of targeted topologies. Alternatively, rigid polynuclear clusters such as $Cu_2(RCOO)_4$ and $Cu_2(Rtz)_2$ can provide more directionality; hence, they can serve as secondary building units (SBUs) [5]. The most frequently used SBUs are $M_2(RCOO)_4L_2$, $M_4(\mu_4\text{-}O)(RCOO)_6$, and $M_3(\mu_3\text{-}O/OH)(RCOO)_6L_3$. When L is a monodentate terminal ligand and the carboxylate is a multitopic ligand, these SBUs can serve as square-planar, octahedral, and trigonal-prismatic nodes, respectively (Fig. 3). On the other hand, if L is a ditopic ligand, the dinuclear $M_2(RCOO)_4$ SBU can serve as a linear (2-connected) linker [27], or a distorted octahedral (6-connected)

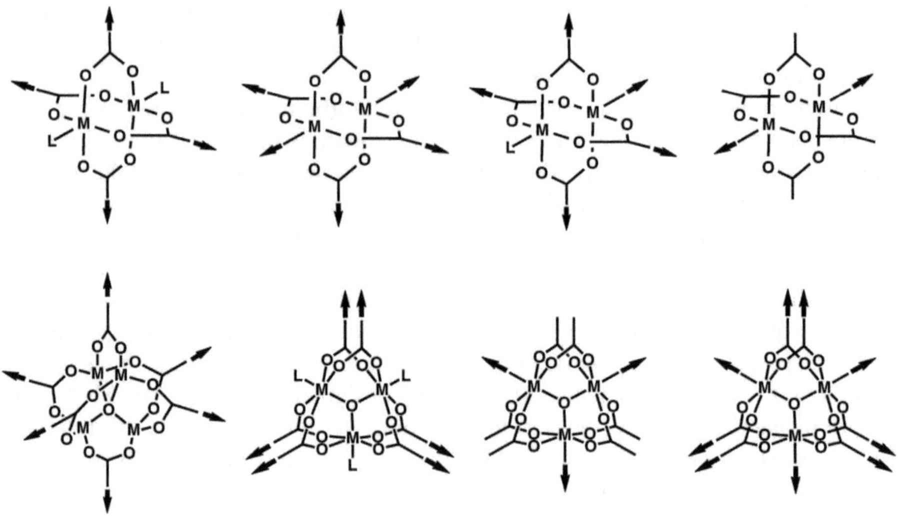

Fig. 3 The $M_2(RCOO)_4$, $M_4(\mu_4\text{-}O)(RCOO)_6$, and $M_3(\mu_3\text{-}O/OH)(RCOO)_6$ SBUs and their representative coordination modes

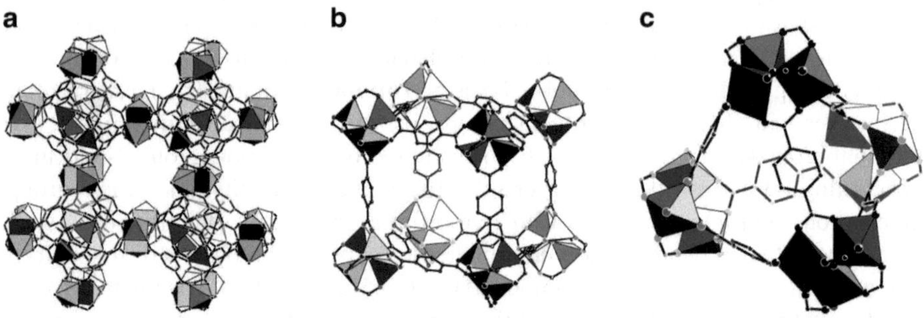

Fig. 4 Structures of (**a**) HKUST-1, (**b**) MOF-5, and (**c**) MIL-101

node [28]. Similarly, the trinuclear $M_3(\mu_3\text{-}O/OH)(RCOO)_6$ SBU can serve as a triangular (3-connected) node [29], or a tricapped trigonal-prismatic (9-connected) node if L is ditopic [30, 31].

For example, Williams et al. reported that solvothermal reaction of $Cu(NO_3)_2$ and benzene-1,3,5-tricarboxylic acid (H_3btc) afforded $[Cu_3(btc)_2(H_2O)_3]$ (HKUST-1) with a highly symmetric porous network with a (3,4)-connected **bor** topology, in which $Cu_2(RCOO)_4(H_2O)_2$ cluster and the btc^{3-} ligand serve as the square-planar and trigonal-planar nodes, respectively (Fig. 4a) [32]. Using expanded tricarboxylate ligands, more porous analogs, sometimes interpenetrated, can be constructed [33, 34]. Yaghi et al. reported a series of **pcu**-type MOFs using the $Zn_4(\mu_4\text{-}O)(RCOO)_6$ cluster and linear dicarboxylate ligands with different lengths and side groups [35], which are isoreticular with the prototypical structure $[Zn_4(\mu_4\text{-}O)(bdc)_3]$ (MOF-5, H_2bdc = benzene-1,4-dicarboxylic acid) (Fig. 4b). Several analogs of this oxo-centered tetranuclear cluster are also known. In a few examples, Zn(II) ions can be fully or partially replaced by Co(II) ions, and the carboxylate groups can be also replaced by pyrazolate groups [36]. Férey et al. developed various 6-connected porous MOFs using the $M_3(\mu_3\text{-O/OH})(RCOO)_6L_3$ clusters. The trigonal-prismatic geometry of $M_3(\mu_3\text{-O/OH})(RCOO)_6L_3$ SBU allows four such clusters to be arranged in a supertetrahedron, which can further interconnect with each other by sharing their vertexes to form zeolitic networks such as MIL-101 (Fig. 4c) [37]. For MOFs based on the $M_2(RCOO)_4L_2$ and $M_3(\mu_3\text{-O/OH})(RCOO)_6L_3$ SBUs, the monodentate terminal ligand L can be removed to generate coordinatively unsaturated metal centers, which can significantly enhance gas sorption affinities. It should be noted that the bonding direction of the carboxylate groups in the $M_3(\mu_3\text{-O/OH})(RCOO)_6L_3$ can change significantly, giving highly flexible frameworks structures when the network topology is suitable [38].

Besides the carboxylate sites, the monodentate coordination sites (or the equatorial sites) can coordinate not only to terminal ligands but also to bipyridyl-type bridging ligands. Generally, the $M_2(RCOO)_4L_2$ clusters form 2D **sql** layers with linear dicarboxylates and different 0D polyhedra with bent dicarboxylates possessing different bridging angles. When linear bipyridyl-type ligands are added, the 2D **sql** $M_2(bdc)_2L_2$ layers can be connected in both sides to form 3D unsymmetrical **pcu** networks with a general composition of $M_2(bdc)_2(bpy)_2$ (bpy = 4,4′-bipyridyl-type ligands). On the other hand, only the terminal ligands on the outer side of the polygons can be substituted by bridging ligands. In this context, the dinuclear SBUs can be regarded as 5-connected nodes (Fig. 3). On the other hand, the polyhedron can be viewed as a super SBU [39]. For example, the $M_2(RCOO)_4L_2$ SBUs can be linked as cubooctahedra and octahedra using bent dicarboxylates with 120° and 90° linking angles, respectively [40, 41], which can be further connected by bipyridyl-type ligands to form **fcu** and **pcu**-type 3D frameworks considering the polyhedra as 12- and 6-connected nodes [39, 42–44]. In this type of porous networks, small cavities are connected to the large channels via the small apertures on the polyhedra.

With three additional connections from the equatorial plane, the $M_3(\mu_3\text{-O/OH})(RCOO)_6L_3$ SBUs can serve as tricapped trigonal-prismatic 9-connected nodes. Using tritopic pyridyldicarboxylate ligands can fulfill the 2:1 molar ratio requirement of carboxylate and pyridyl groups of the trinuclear clusters, which produce a new type of (3,9)-connected topology namely **xmz** (Fig. 5a) [45, 46]. Interestingly, changing the ratio of bridging lengths of the carboxylate and pyridyl ends of the tripodal ligands can drastically alter the bonding direction of the carboxylate groups of the clusters and even generate drastic framework breathing behaviors [47]. Combination of dicarboxylate and pyridylcarboxylate ligands in a 1:2 ratio can also satisfy the coordination requirement of the trinuclear clusters, which give rise to a series of uninodal 9-connected networks with the **ncb** topology (Fig. 5b). The coordination geometries of the clusters in the **ncb** networks are also influenced by the bridging lengths of the ligands. Therefore, the **ncb**-type MOFs can be only synthesized by dicarboxylate-pyridylcarboxylate ligand combinations with suitable bridging length ratios, which was confirmed by geometry calculations and a systematic synthetic trial using a large number of ligand combinations with different bridging lengths [30, 31].

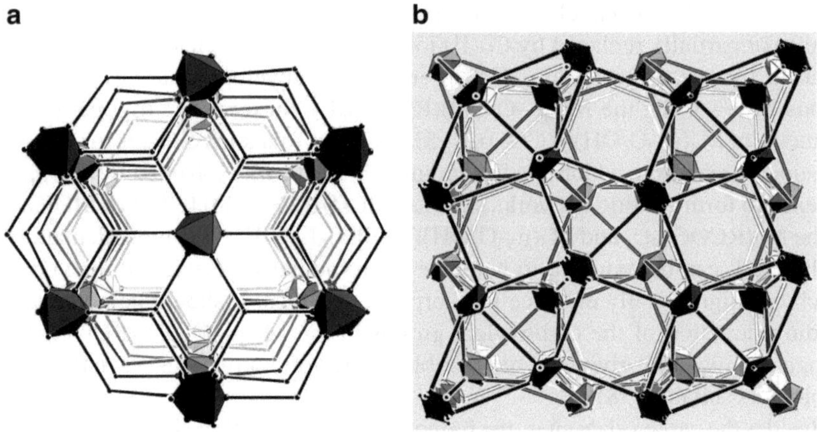

Fig. 5 The (**a**) **xmz** and (**b**) **ncb** topologies

It should be noted that, even when the bridging ligands are very long, no interpenetration was observed for the **ncb**-type MOFs.

2.2 *Framework Interpenetration*

Interpenetration is a common phenomenon for coordination polymers, which is usually regarded as a disadvantage for construction of highly porous frameworks. But in some cases, it can be used to enhance framework stability and increase gas sorption ability (at low pressures) [48–50], or generate interesting properties such as stimuli responsive behaviors [51]. Nevertheless, there are several strategies to avoid framework interpenetration. It is interesting that the **nbo** network seems to prevent network interpenetration because its dual net (i.e., its channel system) is 8-connected **bcu** rather than itself [23]. In other words, interpenetration often occurs for self-dual topologies such as **dia** because crystals tend to contain identical, repeating units. Actually, the non-interpenetrating **ncb** frameworks can be also simplified as **bcu** networks with **nbo** channels regarding the triakistetrahedral supercages as 8-conncted nodes [30, 31]. However, some interpenetrating **nbo** networks have been observed since 2003 [52, 53]. Due to imperfect interpenetration, interpenetrating **nbo** networks are generally porous [54]. Although the self-dual principle refers to perfect interpenetration only, it is still very useful for the selection of topology targets. Compared with the **dia** topology, the more complicated zeolitic topologies are also based on tetrahedral nodes but can hardly show interpenetration. Considering the paramount importance of natural and synthetic zeolites, zeolitic and zeolite-like MOFs have received intensive attention. There are hundreds of known topology types for zeolites [16].

Imidazolate is an exobidentate ligand with bridging angle close to those observed in zeolites. Therefore, it is ideal to combine imidazolate and tetrahedral transition metal ions for zeolitic networks. Such 4-connected coordination networks, such as Zn(II) and

Co(II) imidazolate, have been known for more than 30 years, but they possess neither the zeolitic topology nor porous structures [55, 56]. This situation is just similar the examples mentioned above, i.e., using molecular building blocks mimicking the topology geometry is not enough to construct the desired topology. More in-depth understanding of the topology features should be necessary. By introducing template molecules and/or structure directing agents into the solvothermal reaction system, You and coworkers first synthesized a series of Co(II) imidazolate frameworks with zeolite-like topology [57, 58]. After removal of the guest molecules, most of these structures collapse, this means that these host frameworks with unusual topologies/structures must be stabilized by guest inclusion.

Again, the key for producing subtle structure difference is to use non-coordinated side groups to control the orientation of the imidazolate linkers, which has been comprehensively demonstrated by a large group of low-dimensional coinage metal imidazolates [59–62]. Although a substituent group is generally considered as disadvantageous for generation of porosity, the first porous zeolitic MOF, namely MAF-3, was constructed by Zn(II) and benzimidazolate, which possesses a distorted SOD topology and a porosity of 18% [63]. While the very bulky phenyl ring significantly reduces the porosity and aperture size, it is useful for molecular sieving applications, especially when fabricated as membranes [64]. More porous and regular zeolitic metal imidazolate frameworks, i.e., SOD-$[Zn(mim)_2]$ (MAF-4, Hmim = 2-methylimidazole), ANA-$[Zn(eim)_2]$ (MAF-5, Heim = 2-ethylimidazole) and RHO-$[Zn(eim)_2]$ (MAF-6) were subsequently obtained by using simple alkyl-imidazolates (Fig. 6) [65, 66]. Due to its high porosity, high thermal and chemical stability, and unique pore structure, MAF-4 has served as a prototypical MOF in many studies [18]. Yaghi and coworkers then introduced the high-throughput solvothermal method [67], which enabled the discovery of a variety of new zeolitic and zeolite-like metal imidazolate frameworks (popularly known as zeolitic imidazolate frameworks, ZIFs) with interesting structures and properties [68]. The substituent groups play a very important role in the formation and stability of these zeolitic frameworks, which could be further demonstrated by the successful synthesis of an analogous structure of MAF-4, i.e., SOD-$[Zn(mtz)_2]$ (MAF-7, Hmtz = 3-methyl-1,2,4-triazolate) [69]. Generally, 1,2,4-triazolates tend to use all of the three nitrogen atoms for coordination [70]. But in the case of MAF-7, the methyl group reinforces the formation of an SOD network, leaving an uncoordinated nitrogen atom for guest binding.

There are other strategies for rational construction of non-interpenetrated porous frameworks. For 3D networks, it is obvious that interpenetration can only occur when the porosity of a single network is larger than 50%. Therefore, one may introduce small side groups to reduce the porosity of a single network and yet increase the porosity of the final crystal. Of course, the crystal porosity obtained by this method has an obvious upper limit. Considering that interpenetration of 3D networks must occur in all three dimensions, partial blocking of the coordination network can prevent interpenetration and retain high porosity (Fig. 7).

The pillared-layer structures have been rather widely reported for MOFs, which are not interpenetrated because their relatively dense layers block one of the three dimensions. Based on a suitable layer structure containing coordinatively unsaturated

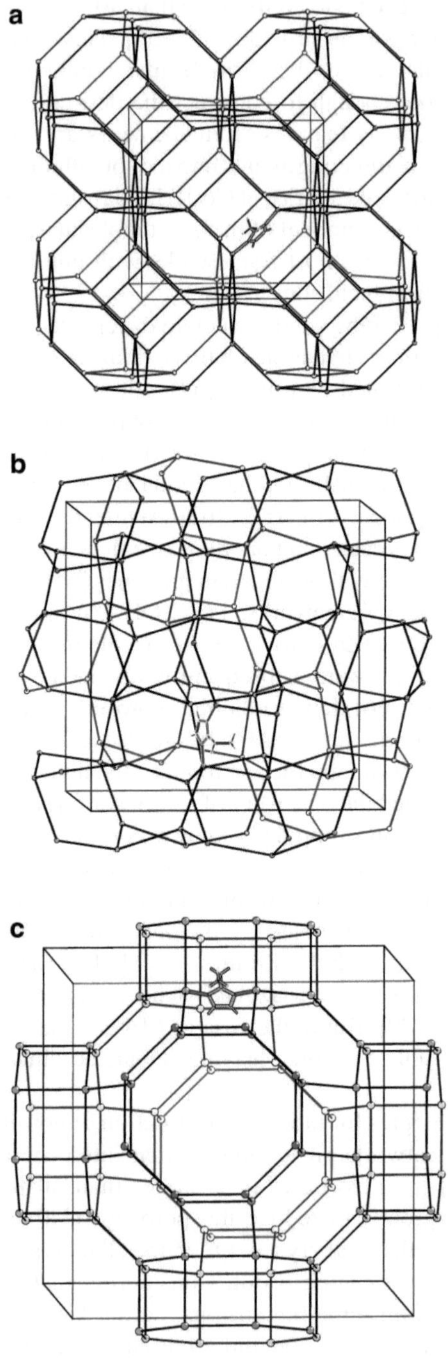

Fig. 6 Framework structures of (**a**) MAF-4, (**b**) MAF-5, and (**c**) MAF-6

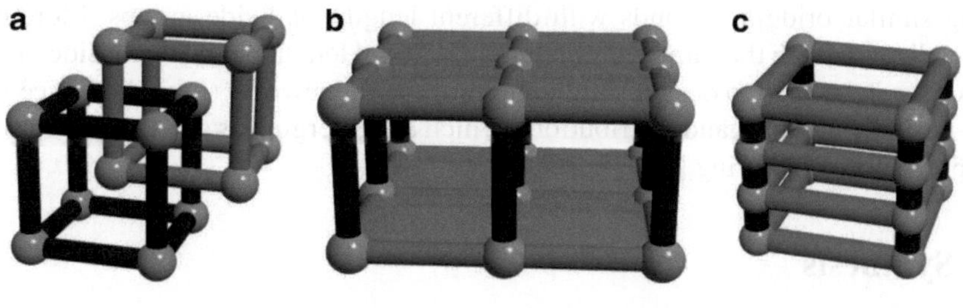

Fig. 7 Control of interpenetration of **pcu**-type structures. (**a**) Normal interpenetration, (**b**) a formal pillared-layer structure consisting of dense layers and long pillars, and (**c**) a non-interpenetrating **pcu** net consisting of porous layers and short pillars

sites, one can change the pillar ligand to obtain MOFs with different porosity and different pore size and surface. The arrangement of pillar ligands is determined by distribution manner of coordinatively unsaturated metal ions on the layer. For example, Kitagawa et al. reported a series of pillared-layer MOFs with a general formula $[Cu_2(pzdc)_2L]$ (CPL-*n*, H_2pzdc = pyrizine-2,3-dicarboxylic acid), in which Cu(II) ions are four-coordinated and bridged by pyrazinedicarboxylate ligands to form dense layers, and these linear bipyridyl-type pillar ligands coordinate with the remained coordination site of Cu(II) and connect the layers into a 3D porous framework. Since the Cu(II) ions on the layer form relatively dense 1D arrays, CPL-*n* possess 1D channels [3].

There are some porous MOFs constructed by porous layers and short pillars, which can be categorized as informal pillared-layer structures which use the short pillars to block two of the three dimensions. For example, when the bipyridyl ligands are relatively long in the $[M_2(bdc)_2(bpy)]$ frameworks with unsymmetrical **pcu** topologies, interpenetration usually occurs [71]. On the other hand, if the pillar ligand is short enough, such as for 1,4-diazabicyclo[2.2.2]octane, no interpenetration can be observed even if the dicarboxylate ligand is very long [72].

The term pillared-layer implies that the 3D structure is hierarchical with the layer formed first and the pillar connects later, but such phenomena can be hardly observed directly since the self-assembly and crystallization of MOFs are generally carried out in a one-pot reaction. Nevertheless, using suitable non-bridging ligands instead of the pillar ligands, the individual single layer structure can be sometimes synthesized. In an occasional case, the hypothetic monolayer and pillared double layer have been isolated during the synthesis of a pillared-layer MOF [73].

In the context of synthesis, the reality is always more complicated than a simple theoretical expectation. The molecular assembly processes are usually influenced by a variety of factors, including the reaction parameters and minor differences in the structures and properties of the metal ion/clusters and the ligands [74]. Therefore, there are still a lot of compounds were unintentionally generated rather than predesigned. However, many compounds have been successfully prepared by the net-based approach or reticular chemistry strategy with either single metal ions or metal clusters as nodes, as well as the pillar-layered strategy. With these strategies, one can construct MOFs with different pore sizes, shapes and surface properties by using similar bridging ligands with different lengths and side groups. Moreover, mixed ligands with the same coordination mode and length but different side groups or cores can be used to design solid-solution-type frameworks (periodic lattice with non-periodic metal/ligand distribution), which are emerging as an effective strategy for pore surface tailoring [69, 75, 76].

3 Synthesis

To allow the self-assembly of metal ions and organic ligands, as well as crystallization, MOFs are mostly synthesized by solution-based methods at different conditions. In the simplest approach, MOF crystals automatically grow in a solution containing the metal salt and organic ligand. When there are solubility problems for the reactants and/or products, other approaches such as diffusion and hydrothermal/solvothermal reactions may be useful, in particular, for growth of large and high-quality crystals for single-crystal diffraction. The reaction method and reaction condition are crucial for the construction of some types of MOFs, especially those containing polynuclear SBUs. For example, solvothermal reactions with amide-based solvents have been widely demonstrated to be very useful for the synthesis of MOFs based on $M_4(\mu_4\text{-}O)(RCOO)_6$ and other types of clusters.

Although some MOF structures were determined by powder diffraction techniques, single crystal specimens with diameter ca. 100 μm and above are usually necessary for structure elucidation of new MOFs. Therefore, the crystallization should be slowed down and undisturbed. On the other hand, to synthesized bulk sample of known structures, the reaction time and scale rather than crystal size are more important. The reaction time of MOF synthesis can be greatly accelerated by increasing reactant concentration, rapid mixing of reactants, heating, microwave irradiation, etc. [77]. Some MOFs can be also rapidly synthesized by mechanochemical reaction of the corresponding metal salt and organic ligand in the presence of very small amount or even without solvent. Nevertheless, soluble metal salts are generally necessary for mechanochemical reactions [78], which produce acid as the by-product and the obtained MOF samples still require solvent washing. In exceptional cases, metal oxide or hydroxide can be used as the reactants, which only generate water as a by-product. For example, some metal imidazolate frameworks have been obtained by solvent/additive-assisted mechanochemical reaction between stoichiometric ZnO and imidazoles at room temperature. A few porous/zeolitic MAFs can be even synthesized without addition of any solvent or additive by simply heating the mixture of $ZnO/Zn(OH)_2$ and the azole ligand [79]. In principle, these mechanochemical and solvent-free reaction methods can greatly increase the space-time-yield and beneficial for large-scale production, but difficult to ensure the product purity because impurity is always contained by the reactants. Alternatively, higher-purity MAFs can be obtained by aqueous ammonia-assisted reactions, in which ammonia increases the solubility of metal oxides and helps the deprotonation of weakly acidic azoles [80].

While many structure design strategies have been developed, none of which can guarantee that there is only one possible self-assembly result with a given set of reactants. Note that solvent and counter ions always participate in the reaction system and often involve in the final products. When the chemical composition of the coordination network is fixed, the obtained product would be limited to a few supramolecular isomers. Energetically, porous networks are generally less stable than nonporous ones. Therefore, as-synthesized MOF crystals always contain solvent and/or template molecules, which help stabilizing the porous structures. For MOFs containing large channels and small guest molecules, it is not easy to judge if the guest molecules serve as templates.

In some cases, the template effect is crucial and can be clearly demonstrated by the crystal structures. For example, Kitagawa et al. showed that reaction of $Zn(NO_3)_2$, Li(TCNQ) (TCNQ = 7,7,8,8-tetracyano-*p*-quinodimethane) and bpy in pure methanol and mixed methanol/benzene gave two MOFs with very different structures, namely [Zn(TCNQ)(bpy)]·6MeOH and [Zn(dTCNQ)(bpy)]·1.5benzene [81, 82]. The $TCNQ^-$ anion disproportionates and dimerizes to form the $TCNQ^{2-}$ and $dTCNQ^{2-}$ dianions without and with benzene addition, respectively (Fig. 8). These dianion ligands contain excess electrons which can form strong charge transfer-type interactions with aromatic molecules. The benzene molecules are embedded in the small cavities composed of the $dTCNQ^{2-}$ ligands. In this context, the template molecule can even influence the redox equilibrium of the reaction system, resulting in unusual in situ ligand transformation during MOF synthesis.

The in situ ligand reactions have been observed more frequently during solvothermal reactions, in which the high reaction temperature and catalytic action of metal ions may promote organic reactions that are infeasible at common conditions. Actually, many in-situ metal/ligand reactions have been reported to be useful for synthesis of coordination polymers, such as hydrolysis of carboxylate esters, organic nitriles, and aldehydes into the corresponding carboxylates, cleavage of acetonitrile/ethylene carbon–carbon bonds and 1,3,4-oxadiazole carbon–nitrogen/carbon–oxygen bonds, cleavage and formation of disulfide bonds, substitution of aromatic groups, decarboxylation of aromatic carboxylates, cycloaddition of organic nitriles with azide and ammonia, and so on (Fig. 8) [83, 84]. For example, in situ solvothermal generation of carboxylate groups by hydrolysis of nitrile or ester groups is an effective synthetic method to obtain the non-centrosymmetrical (or nonlinear optical active) coordination polymers, which are not easily formed by conventional reactions of carboxylic acids and metal ions [22]. In situ solvothermal reactions of iron powder with organic carboxylic acid are also critical for the syntheses of magnetically interesting Fe(II) and mixed-valence Fe(II,III) coordination polymers [85, 86], and some mixed-ligand coordination polymers, which are hardly to be prepared by direct mixing of the related ligand and metal ions, could possibly be achieved by decomposition of an organic molecule into two different bridging ligands [87].

It is also noteworthy that MOFs, although generally regarded to be much instable than conventional adsorbents, can undergo covalent post-synthetic modification or metal/ligand exchange [88, 89]. Nevertheless, hydrothermal/solvothermal reactions are important for searching new, interesting MOFs and growing large crystals for structural characterization by precise single-crystal X-ray diffraction. However, for practical applications, large-scale and rapid synthesis is also a very important objective. Therefore, conventional reactions by direct mixing of the reactants, as well as microwave-assistant and mechanochemical reactions are very attractive.

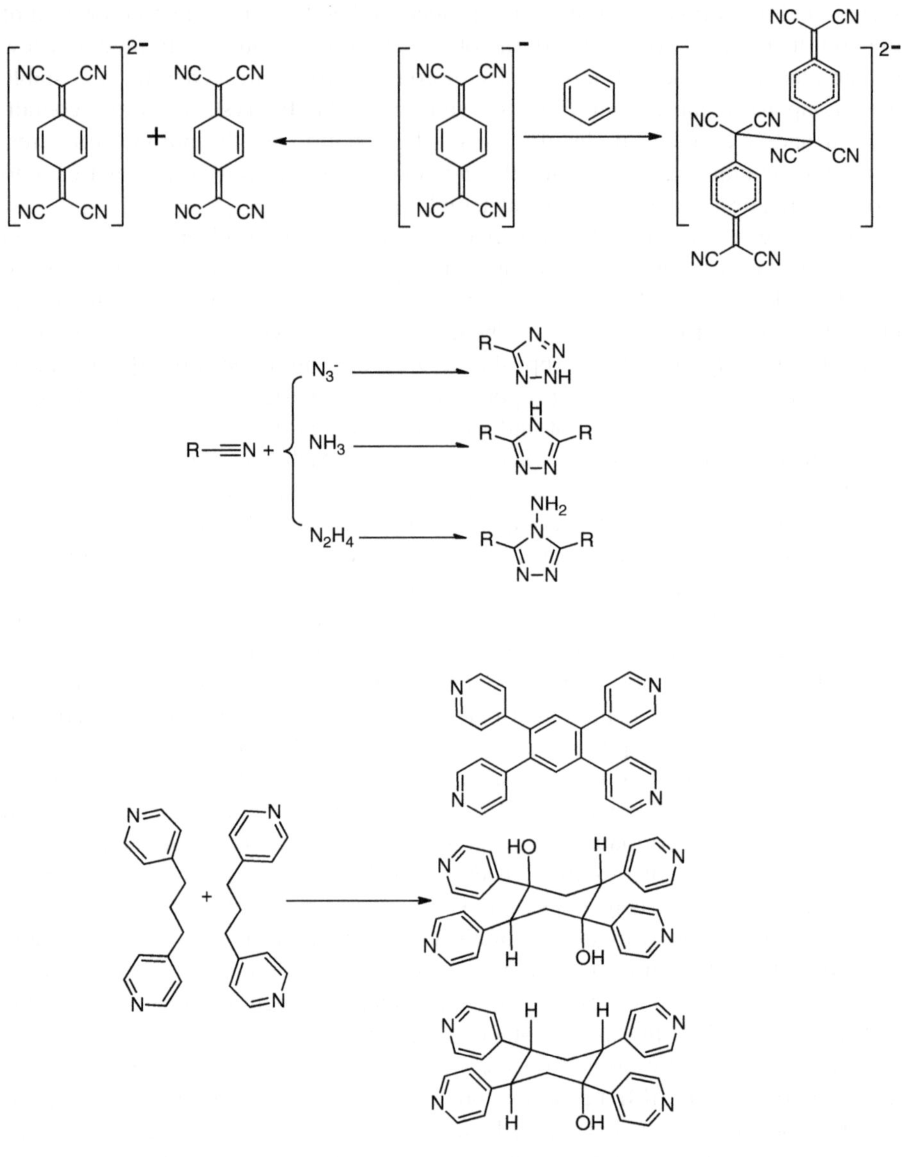

Fig. 8 Representative in situ organic reactions occurred during MOF synthesis

4 Properties and Applications

By virtue of the 3D periodic structures, the porous structures of MOFs can be straightforwardly visualized from the crystal structures, which is very important for understanding the structure–property relationship. However, the host framework may not retain after guest removal, because coordination bonds are relatively weak compared with covalent bonds in traditional adsorbents. Many MOFs collapse after guest removal and heat treatment, meaning that there are only pseudo or virtual porosity in the as-synthesized crystals. Therefore, rigid frameworks with high thermal stability have been the main research target [90]. So far, many MOFs can be stable up to 300–400°C and a few to 500°C, which are high enough for many applications. Actually, the chemical stability of MOFs is a more serious problem. Most carboxylate-based MOFs are unstable in water or moist air. In contrast, MAFs especially those based on diazolates (imidazolates and pyrazolates) have demonstrated high thermal and chemical stabilities, which can be attributed to their relatively strong metal-azolate coordination bonds and chemically inert pore surfaces.

To confirm the permanent porosity of the sample, a gas sorption isotherm is necessary, which is commonly carried out by nitrogen at 77 K. Argon with smaller size and spherical molecular shape can measure pore surface area and pore volume more precisely. When the pore size is too small to be entered by nitrogen, carbon dioxide and hydrogen with smaller kinetic diameters can be used instead. The gas sorption behavior is determined by the adsorbate, the temperature and pressure, and the structural characteristics of adsorbent such as pore size and shape, surface area, and pore volume. By virtue of the diversified and unique structures, MOFs have demonstrated many interesting properties potentially useful for practical applications.

4.1 Framework Flexibility

While rigid frameworks are associated with high stability and usefulness for practical applications, many MOFs can change their structures under external stimuli, because their host frameworks consist of relatively weak bonds, rotatable single bonds, weak supramolecular interaction between multiple networks, etc. [9, 91, 92]. For example, even for the seemingly very rigid structure MOF-5, the phenyl ring of the bdc^{2-} ligand is rotatable [93], although the overall crystal lattice is not distorted. More commonly, rotation of molecular building blocks around the metal-ligand bonds or single bonds in the organic ligands can give rise to framework distortion and crystal symmetry change. Flexible coordination networks can also lose the long-range order to form amorphous or quasi-amorphous phases. While MOFs with the pseudo-porosity which irreversibly turn amorphous after guest removal, flexible MOFs show reversible structural transformations either in a crystal to crystal or in a crystal to amorphous manner. For example, 3,3′,5,5′-tetramethyl-4,4′-bipyrazolate (bpz^{2-}) has been used as flexible ditopic ligand to link trinuclear Ag_3 and Cu_3 clusters as 3-connected porous frameworks [94–96]. The two pyrazolate rings are rotatable around their connection (a C–C single bond), which possess a dihedral angle freely changeable between 50° and 90°. Therefore, the porous MOF [Ag_2(bpz)] demonstrated a variety of crystal symmetry after inclusion of different guests. In specific guest loading, a single-crystal specimen may lose its diffraction ability although its shape and transparency are not significantly changed, indicating the formation of a highly distorted, non-periodic network [95, 96].

To monitor the structure and crystallinity change of MOFs, powder X-ray diffraction is a convenient technique. More precise host–guest structural information could be obtained by synchrotron diffraction and single-crystal X-ray diffraction [97, 98]. Combination of other in-situ spectroscopic techniques can give more insightful structural information [99], especially for non-periodic structural alterations.

Flexible MOFs can show very different types of framework flexibilities such as gliding of 2D layers [19], shifting of interpenetrating 3D networks [100], huge breathing of a single 3D network due to changeable bonding direction of organic ligands [38]. The structural transformation of host framework is always associated with energy change, which is generally induced by chemical stimuli, i.e., the sorption, desorption, and exchange of guest molecules. This common kind of framework flexibility usually shows abnormal guest sorption isotherms. Framework flexibility induced by physical stimuli such as light and heat has been rarely reported. For example, MAF-2 processes an **nbo-a**-type host framework, which defines a **bcu**-type channel system with large cavities interconnected by small apertures. Interestingly, the apertures are blocked by the obviously flexible ethyl groups, and N_2 adsorption was observed at 195 K rather than at 77 K. Further, single-crystal structures of the guest-free and N_2 loaded phases showed that there is only one thermodynamically stable state for the host framework, indicating that the ethyl group is flexible with very low energy change and energy barrier, so that its motion is determined mainly by heat instead of guest. After adsorption of different solvent molecules, MAF-2 can also distort its metal triazolate framework to form cubic, trigonal, and triclinic crystal symmetry (Fig. 9) [26]. Obviously, the ability of flexible MOFs to change their host structures by chemical and physical stimuli is a unique characteristic compared with conventional adsorbents, which may be used to produce novel and useful properties.

4.2 Gas Storage

With appropriate framework design, many MOFs possess surface area and pore volume much larger than conventional adsorbents such as activated carbons and zeolites, which imply that MOFs are promising materials for gas storage. Nevertheless, the adsorbent is potentially useful only when its adsorption amount is larger than that of an empty container. For example, MOFs with large pore volumes can always adsorb very large amounts of nitrogen gas at 77 K, but it is not useful because an empty container can store more liquid nitrogen at 77 K, or a gas cylinder working at above 100 bar can easily store much gaseous nitrogen at room temperature. Similarly, MOFs may adsorb more CO_2 than an empty cylinder at room temperature at the low pressures, but they are not considered as good CO_2 storage medium because CO_2 can be liquefied easily without excess pressurization.

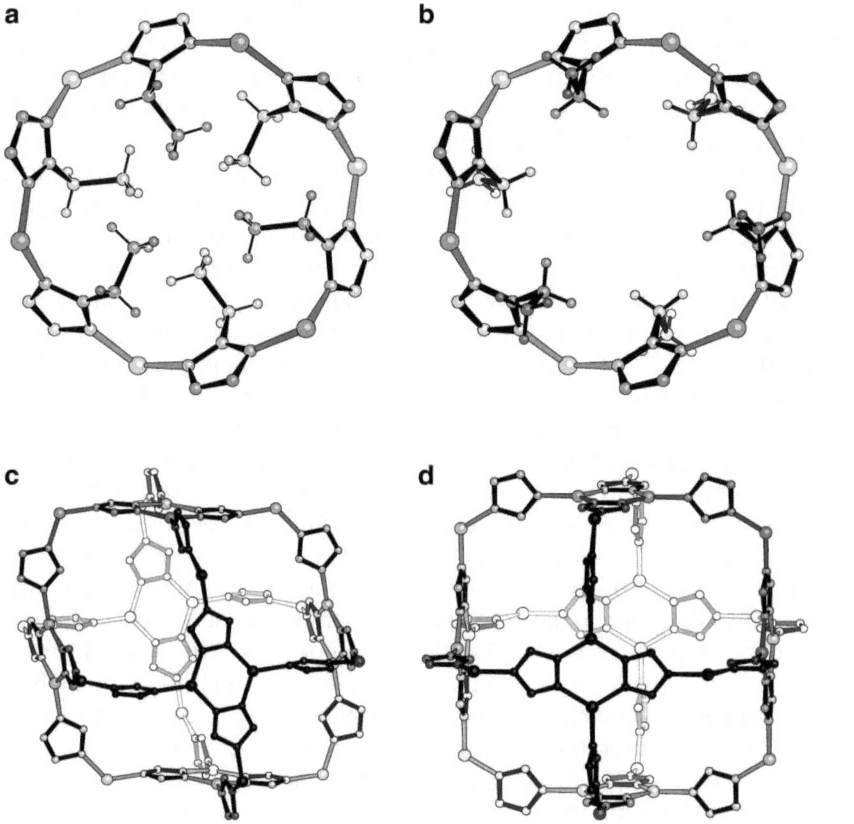

Fig. 9 Two kinds of framework flexibility observed in MAF-2. (**a**, **b**) Temperature-induced closed and open states of the ethyl blocked aperture. (**c**, **d**) Guest-induced distortion of the Cu-triazolate framework

Therefore, MOFs are generally considered as the storage medium for gases such as hydrogen and methane that are difficult to compress or liquefy.

Hydrogen is a clean and high-energy fuel, but it is very difficult to store or deliver because of its extremely low boiling point. Various strategies have been studied for hydrogen storage, but none has met the practical requirements yet. For example, chemical absorption can provide very large storage capacity but hydrogen can be hardly released. Alternatively, physical adsorption of hydrogen in porous materials such as MOFs is very fast [101, 102]. By virtue of the ultrahigh surface areas and pore volumes, some MOFs have achieved the DOE (Department of Energy, U.S.A.) targets for hydrogen storage, i.e., 6.5 wt% and 45 g L^{-1}. Note that the density of liquid hydrogen is only ca. 70 g L^{-1}. However, these data can be only obtained at very low temperatures such as 77 K, because the adsorption enthalpy of physical adsorption is too low. There are different strategies to enhance the adsorption affinity of MOFs. Pore size is a critical parameter that determines the adsorption affinity. However, very small pore size is not likely compatible with the very large surface area and pore volume required for practical applications. Incorporation of coordinatively unsaturated metal centers on the pore surface can also effectively enhance adsorption affinity, but high-concentration metal ions would generally increase the framework density.

As a main component in natural gases, methane is also an important fuel. The boiling point of methane is very low but much higher than that of hydrogen. Therefore, methane can be easier adsorbed by porous materials. A few MOFs have demonstrated large methane storage capacity exceeded the DOE target, i.e., 180 v/v at room temperature and 35 bar. For example, Zhou and coworkers designed and synthesized a highly porous coordination network PCN-14, in which the $Cu_2(RCOO)_4$ clusters are connected by anthracene tetracarboxylate ligands into an **nbo** topology [103]. PCN-14 possesses a large Langmuir surface area over 2,100 $m^2\ g^{-1}$ and a large pore volume over 0.8 $cm^3\ g^{-1}$, as well as nanocage-type pore and large aromatic rings on the pore surface, which give rise to methane uptake of 230 v/v and enthalpy of 30 kJ mol^{-1} at 290 K and 35 bar.

While increasing the adsorption enthalpy or affinity has been the main strategy for improving the gas storage performance, storage refers to not only deposit but also withdraw, i.e., adsorption and desorption of an adsorbate. Actually, the effective storage capacity equals to the difference of adsorption amounts at two temperature/pressure points of interest rather than the adsorption amount at a single temperature/pressure point. When the adsorption enthalpy is too high, desorption becomes difficult although adsorption amount is high. It has been calculated that, assuming a type-I hydrogen adsorption isotherm working between 1.5 and 35 bar at 298 K for a typical storage system, the maximum storage capacity can be obtained by an adsorption enthalpy of 15.1 kJ mol^{-1}[104]. Actually, the type-I isotherm is very steep at the lowest pressure ranges, which means very large amounts of adsorbates cannot be desorbed. For example, the highly porous MOF-177 has not only a very high hydrogen uptake of 48.3 g L^{-1} at 77 K and 70 atm, but also a relatively high uptake of 8 g L^{-1} at 1.5 bar, meaning that its useful storage capacity is about 40 g L^{-1} [105].

In special cases, the concept of useful storage capacity is more crucial. For example, acetylene is highly unstable, and pure acetylene cannot be compressed above 2 bar. Therefore, gas cylinder can only provide very limited storage capacity because the working pressure difference is small. Although some microporous adsorbents such as zeolites and MOFs can adsorb much more acetylene than for an empty gas cylinder [106], their adsorption affinity is too strong, which prevent acetylene desorption above 1 bar. While a normal type-I sorption isotherm tends to retain much guest at low pressure, a type III or V isotherm can provide more useful storage capacity, but these isotherms are generally associated with low saturated adsorption amounts and weak adsorption affinity (Fig. 10). Therefore, the adsorption affinity should be lowered or the isotherm shape should be modified to maximize the uptake difference between the lower and upper working temperature for pure acetylene, which can be achieved by an MOF with suitable pore surface structure and framework flexibility [107].

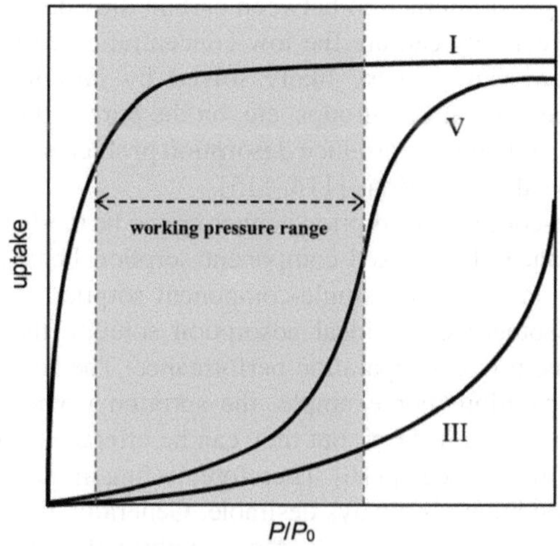

Fig. 10 Comparison of the usable storage capacity of Type I, III, and V isotherms

4.3 Separation and Enrichment

The sorption behaviors of different adsorbates are always different, which can be used to separate or enrich one of the adsorbates from the mixtures [108, 109]. Unlike storage, separation does not require large surface area and pore volume. Alternatively, small pore size is commonly beneficial for high separation performance. Adsorbents can separate different adsorbates in several mechanisms. The most effective one is the molecular sieving effect, which only adsorbs the smaller molecules and excludes the larger molecules completely, giving very high separation ratio. Crystalline MOFs are highly ordered frameworks as for zeolites, which can be also engineered to have suitable pore sizes and shapes [110]. Further, flexible MOFs can selectively change their pore sizes. To utilize the molecular sieving properties, adsorbents may be used in different manners. The most straightforward and economic method is to fabricate an adsorbent-based membrane. Gas sieving membranes have been successfully fabricated with several types of MOFs such as MAF-3, MAF-4, and HKUST-1 [64, 111, 112].

More commonly, two kinds of molecules are separated by the different adsorption affinity or adsorption speed. For example, carbon dioxide is regarded as a main source of green house effect, and separation of carbon dioxide from fuel gases has attracted intensive attention recently [113]. Although chemical absorption of carbon dioxide by amine-based solvents is very effective, the regeneration process requires much energy, and the volatile, corrosive amine vapor is also a serious problem. Physical adsorption is a viable approach to capture carbon dioxide. The boiling point and quadruple moment of carbon dioxide is much higher than those of nitrogen and oxygen in fuel gasses, which means carbon dioxide can be selectively adsorbed.

However, the host–guest interaction between carbon dioxide and MOFs are generally too weak to effectively capture the low-concentration carbon dioxide. Nevertheless, this problem can be very likely solved by introducing coordinatively unsaturated metal centers, amine groups, etc. on the pore surfaces. Of course, high adsorption enthalpy will always produce desorption problems, which may be solved by using multiple weak interactions [114, 115].

Generally, single-component sorption isotherms can be used to judge the potential separation performance. The mixed-component sorption behavior is not a simple mathematical addition of several single-component sorption isotherms. With only single-component isotherms, the ideal adsorption solution theory (IAST) can be applied to predict the mixture separation performance. The sorption kinetics is also very crucial for separation. For example, the sorption isotherms of propane and propene are very similar in MAF-4, but they can be effectively separated according to their different diffusion speed [116]. Therefore, testing the separation performance in more practical conditions is always desirable. Generally, the practical separation performance of an MOF can be tested by the breakthrough experiment, in which the mixture passes through the MOF filled column, and the time-dependent concentrations of adsorbates are recorded. Some practical devices have been fabricated by MOF crystals or thin films. For example, Chen et al. synthesized the twofold interpenetrating framework [$Zn_2(bdc)_2(bpy)$] (MOF-508), in which the narrow channel (4.0 × 4.0 Å^2) is slightly larger than that of a methane molecule (3.8 × 3.8 Å^2). A gas chromatographic (GC) column was packed with crystals of guest-free MOF-508, which can efficiently separate linear and branched alkanes based on a molecular sieving effect [71]. Using nanocrystals of MAF-4, Yan et al. fabricated capillary GC columns, which showed high-resolution chromatographic separation of linear alkanes [117].

The high porosity and structural regularity of MOFs are suitable for molecular enrichment and separation. A few MOFs have been used for solid-phase microextraction (SPME), in which porous organic polymers are traditionally employed to coat the extracting fiber. The first MOF-coated SPME fiber was reported by Yan et al. using HKUST-1, which showed large enrichment and low limits of detection for benzene homologues, but the extraction efficiency is largely reduced at humidity over 30% due to the poor water stability of HKUST-1 [118]. They also fabricated highly water-stable SPME fibers based on MAF-4 and MAF-3. Due to the very small aperture sizes of MAF-4 (3.4 Å) and MAF-3 (2.9 Å), the former showed superior selectivity for *n*-alkanes over branched alkanes, while the latter adsorbed neither, and larger molecules such as benzene homologues can be only adsorbed on the outer surfaces [119]. Using 4-(3,5-dimethylpyrazol-4-yl) benzoic acid (H_2mpba) as a bifunctional ligand, Zhang et al. synthesized a highly stable and porous MOF [Zn(mpba)] (MAF-X8), which possesses large and regular 1D channels with a relatively hydrophobic pore surface. MAF-X8 can be directly grown on the surface of stainless-steel fiber to form thin layers with channels exposed, which showed not only extraction efficiency for benzene homologues much higher than conventional organic polymer-coated SPME fibers but also superior selectivity even in the presence of large amounts of phenols. As the benzene homologues and phenols have similar molecular sizes, the molecular sieving effect was attributed to the difference of diffusion speed, which was proven by time-dependent extraction experiments and molecular dynamic calculations [120].

4.4 Molecular Sensing

The unique coordination structures of MOFs consisting of various metal ions/clusters and organic ligands give rise to not only unique sorption behaviors but also rich optic, electric, and magnetic properties [121, 122]. Frequently, these physical properties are also influenced by the guest species or host–guest interactions. Further, guest adsorption, desorption, and exchange can change the host framework structures of flexible MOFs, which alter the physical properties more significantly. In turn, the species and concentration of guest molecules can be deduced by the change of structures and physical properties of MOF crystals [123]. For example, Chen et al. showed that a microporous MOF $[KCo_7(OH)_3(ipa)_6(H_2O)_3]$ (H_2ipa = isophthalic acid) can reversibly release/reload the water molecule coordinated with the Co(II) ions in a single-crystal to single-crystal transformation manner, during which some of the six-coordinated Co(II) ions are changed into pseudo five- and four-coordinated ones. The color of the crystals is reversibly changed between red and purple, being similar to that of commonly used silica gels. The change of coordination geometry of the Co(II) ions also alters the magnetic properties between superparamagnetism and paramagetism [124]. Also, spin crossover, which is based on the high and low spin transition of single metal ions and very sensitive to the coordination geometry, electronic structure of organic ligand, and guest perturbation of crystal lattice, can be used as mechanism for guest sensing [125–127].

Compared with other physical properties, the change of color or photoluminescence can be more easily identified and measured. Color change of MOFs can be originated from the intrinsic properties of the chromophore involved [128], or from the crystallite/film structures of the devices [129, 130]. The rich photophysical properties of metal complexes in both solution and solid states have been widely studied for a long time. The luminophors are regularly immobilized in the crystal lattices of MOFs with high concentration. The highly porous structures of MOFs can enhance the interaction between the luminophors and guest molecules in the confined space. A variety of guest species can be detected by photoluminescent MOFs [122, 131]. For example, Chen et al. showed that small molecules such as dimethyl formamide, acetone, and ethanol exhibit different enhancing and quenching effects on the photoluminescence of a microporous MOF [Eu(btc)] by binding to the Eu(III) ions [132]. Li et al. reported microporous MOFs with high quenching efficiencies for organic compounds containing nitro groups, which can be used to detect chemical explosives [133]. Lin et al. showed that the classic phosphorescent precious metal-based complexes can be decorated as metallo-ligands to construct MOFs highly sensitive to molecular oxygen [134].

The interaction between luminophors can significantly change the luminescence properties such as emission wavelengths and lifetimes. In solution, the luminophor–luminophor interactions are generally controlled by the concentration. In MOFs, the average separation between adjacent luminophors are much shorter than those in solution, but the self-quenching effect or excimer formation can be avoided because the luminophors are fixed by the coordination network. Nevertheless, the luminophor–luminophor interaction can be rationally changed by framework flexibility or doping. For example, 1*H*-imidazo[4,5-*f*][1, 10]phenanthroline (Hip) is a large π-conjugated molecule displaying strong blue photoluminescence in the discrete state (such as in dilute solution), but it is non-luminescent in the solid state due to extensive intermolecular interactions and energy transfer. Zhang et al. reported a 3D flexible porous framework $[Zn_7(ip)_{12}](OH)_2$, in which the ligand–ligand interaction can be gradually changed by guest-induced framework distortion, giving large change of emission color from blue to orange [135].

MOFs also showed other types of photonic applications, which are discussed in other chapters of this book. Besides the interesting properties and applications discussed above, MOFs can serve as catalysts and catalyst supporters [136–138], as reaction containers for small organic molecules and polymers [139, 140], as well as biomedicine [141] and second-order nonlinear optics [142].

5 Conclusions

With the development in molecular design, such as the zeolitic and metal cluster-based net strategies, and synthetic approaches in the last two decades, a large number of very interesting MOFs have been uncovered. The considerable efforts have demonstrated that MOFs have not only designable and diverse framework structures with different pore sizes and shapes but also modifiable and flexible frameworks. Such characteristics provide better molecular recognition abilities associated with the unique framework flexibilities and host–guest molecular interactions, hence exhibit higher performance than other porous materials. Meanwhile, the recent progress in fabrication of this kind of materials into devices significantly shortens the road to practical applications.

We can anticipate that, by further efforts and, in particularly collaborating with engineers, researchers will be able to design and prepare more practically applicable MOF-based materials with unique pore surface, sizes, shapes, and framework flexibilities for excellent molecular recognition abilities and higher performance than other porous materials, in adsorption, separation, sensing, catalysis and other aspects, and to finely modify and fabricate these materials into practical devices.

Acknowledgment This work was supported by the "973 Project" (2012CB821706) and NSFC (21290173, 21225105 and 21121061), Science and Technology Department of Guangdong Province (S2012030006240).

References

1. Hoskins BF, Robson R (1989) J Am Chem Soc 111:5962
2. Hoskins BF, Robson R (1990) J Am Chem Soc 112(1546):1554
3. Kondo M, Okubo T, Asami A, Noro S, Yoshitomi T, Kitagawa S, Ishii T, Matsuzaka H, Seki K (1999) Angew Chem Int Ed 38:140
4. Li H, Eddaoudi M, O'Keeffe M, Yaghi OM (1999) Nature 402:276
5. Tranchemontagne DJ, Mendoza-Cortes JL, O'Keeffe M, Yaghi OM (2009) Chem Soc Rev 38:1257
6. Batten SR, Champness NR, Chen XM, Garcia-Martinez J, Kitagawa S, Ohrstrom L, O'Keeffe M, Suh MP, Reedijk J (2012) CrystEngComm 14:3001
7. Ferey G (2008) Chem Soc Rev 37:191
8. Czaja AU, Trukhan N, Muller U (2009) Chem Soc Rev 38:1284
9. Horike S, Shimomura S, Kitagawa S (2009) Nat Chem 1:695
10. Robson R (2000) J Chem Soc Dalton Trans 3735
11. Yaghi OM, O'Keeffe M, Ockwig NW, Chae HK, Eddaoudi M, Kim J (2003) Nature 423:705
12. Wells AF (1977) Three-dimensional nets and polyhedra. Wiley, New York
13. Reticular chemistry structure resource (RCSR). http://rcsr.anu.edu.au/
14. O'Keeffe M, Peskov MA, Ramsden SJ, Yaghi OM (2008) Acc Chem Res 41:1782
15. Blatov VA, O'Keeffe M, Proserpio DM (2010) CrystEngComm 12:44
16. IZA structure commission. http://izasc.ethz.ch/fmi/xsl/IZA-SC/ft.xsl
17. Zhang J-P, Huang X-C, Chen X-M (2009) Chem Soc Rev 38:2385
18. Zhang JP, Zhang YB, Lin JB, Chen XM (2012) Chem Rev 112:1001
19. Biradha K, Hongo Y, Fujita M (2002) Angew Chem Int Ed 41:3395
20. Fletcher AJ, Cussen EJ, Bradshaw D, Rosseinsky MJ, Thomas KM (2004) J Am Chem Soc 126:9750
21. Noro S, Kitagawa S, Kondo M, Seki K (2000) Angew Chem Int Ed 39:2082
22. Evans OR, Lin WB (2002) Acc Chem Res 35:511
23. Eddaoudi M, Kim J, O'Keeffe M, Yaghi OM (2002) J Am Chem Soc 124:376
24. Zhang J-P, Zheng S-L, Huang X-C, Chen X-M (2004) Angew Chem Int Ed 43:206
25. Zhang JP, Lin YY, Huang XC, Chen XM (2005) J Am Chem Soc 127:5495
26. Zhang J-P, Chen X-M (2008) J Am Chem Soc 130:6010
27. Batten SR, Hoskins BF, Moubaraki B, Murray KS, Robson R (2000) Chem Commun 1095
28. Dybtsev DN, Chun H, Kim K (2004) Angew Chem Int Ed 43:5033
29. Zheng YZ, Tong ML, Xue W, Zhang WX, Chen XM, Grandjean F, Long GJ (2007) Angew Chem Int Ed 46:6076
30. Zhang YB, Zhang WX, Feng FY, Zhang JP, Chen XM (2009) Angew Chem Int Ed 48:5287
31. Zhang YB, Zhou HL, Lin RB, Zhang C, Lin JB, Zhang JP, Chen XM (2012) Nat Comm 3:642
32. Chui SSY, Lo SMF, Charmant JPH, Orpen AG, Williams ID (1999) Science 283:1148
33. Chen BL, Eddaoudi M, Hyde ST, O'Keeffe M, Yaghi OM (2001) Science 291:1021
34. Wang XS, Ma SQ, Yuan DQ, Yoon JW, Hwang YK, Chang JS, Wang XP, Jorgensen MR, Chen YS, Zhou HC (2009) Inorg Chem 48:7519
35. Eddaoudi M, Kim J, Rosi N, Vodak D, Wachter J, O'Keeffe M, Yaghi OM (2002) Science 295:469
36. Hou L, Lin YY, Chen XM (2008) Inorg Chem 47:1346
37. Ferėy G, Mellot-Draznieks C, Serre C, Millange F, Dutour J, Surble S, Margiolaki I (2005) Science 309:2040
38. Serre C, Mellot-Draznieks C, Surble S, Audebrand N, Filinchuk Y, Ferey G (2007) Science 315:1828
39. Perry JJ, Perman JA, Zaworotko MJ (2009) Chem Soc Rev 38:1400
40. Eddaoudi M, Kim J, Wachter JB, Chae HK, O'Keeffe M, Yaghi OM (2001) J Am Chem Soc 123:4368
41. Ni Z, Yassar A, Antoun T, Yaghi OM (2005) J Am Chem Soc 127:12752

42. Chun H (2008) J Am Chem Soc 130:800
43. Chun H, Jung HJ, Seo JW (2009) Inorg Chem 48:2043
44. Li JR, Timmons DJ, Zhou HC (2009) J Am Chem Soc 131:6368
45. Zhang XM, Zheng YZ, Li CR, Zhang WX, Chen XM (2007) Cryst Growth Des 7:980
46. Jia JH, Lin X, Wilson C, Blake AJ, Champness NR, Hubberstey P, Walker G, Cussen EJ, Schroder M (2007) Chem Commun 840
47. Wei Y-S, Chen K-J, Liao P-Q, Zhu B-Y, Lin R-B, Zhou H-L, Wang B-Y, Xue W, Zhang J-P, Chen X-M (2013) Chem Sci 4:1539
48. Sun DF, Ma SQ, Ke YX, Collins DJ, Zhou HC (2006) J Am Chem Soc 128:3896
49. Ma SQ, Sun DF, Ambrogio M, Fillinger JA, Parkin S, Zhou HC (2007) J Am Chem Soc 129:1858
50. Ma SQ, Eckert J, Forster PM, Yoon JW, Hwang YK, Chang JS, Collier CD, Parise JB, Zhou HC (2008) J Am Chem Soc 130:15896
51. Takashima Y, Martinez VM, Furukawa S, Kondo M, Shimomura S, Uehara H, Nakahama M, Sugimoto K, Kitagawa S (2011) Nat Comm 2:168
52. Chen BL, Fronczek FR, Maverick AW (2003) Chem Commun 2166
53. Bu XH, Tong ML, Chang HC, Kitagawa S, Batten SR (2004) Angew Chem Int Ed 43:192
54. Lin J-B, Zhang J-P, Chen X-M (2010) J Am Chem Soc 132:6654
55. Sturm M, Brandl F, Engel D, Hoppe W (1975) Acta Crystallogr B 31:2369
56. Lehnert R, Seel F (1980) Z Anorg Allg Chem 464:187
57. Tian YQ, Cai CX, Ji Y, You XZ, Peng SM, Lee GH (2002) Angew Chem Int Ed 41:1384
58. Tian YQ, Cai CX, Ren XM, Duan CY, Xu Y, Gao S, You XZ (2003) Chem Eur J 9:5673
59. Huang XC, Zhang JP, Chen XM (2004) J Am Chem Soc 126:13218
60. Huang X-C, Zhang J-P, Lin Y-Y, Chen X-M (2005) Chem Commun 2232
61. Huang X-C, Zhang J-P, Chen X-M (2006) Cryst Growth Des 6:1194
62. Zhang JP, Qi XL, He CT, Wang Y, Chen XM (2011) Chem Commun 47:4156
63. Huang X-C, Zhang J-P, Chen X-M (2003) Chin Sci Bull 48:1531
64. Li YS, Liang FY, Bux H, Feldhoff A, Yang WS, Caro J (2010) Angew Chem Int Ed 49:548
65. Huang X-C (2004) PhD Thesis thesis, Sun Yat-Sen University, Guangzhou
66. Huang XC, Lin YY, Zhang JP, Chen XM (2006) Angew Chem Int Ed 45:1557
67. Banerjee R, Phan A, Wang B, Knobler C, Furukawa H, O'Keeffe M, Yaghi OM (2008) Science 319:939
68. Phan A, Doonan CJ, Uribe-Romo FJ, Knobler CB, O'Keeffe M, Yaghi OM (2010) Acc Chem Res 43:58
69. Zhang J-P, Zhu A-X, Lin R-B, Qi X-L, Chen X-M (2011) Adv Mater 23:1268
70. Zhang J-P, Chen X-M (2006) Chem Commun 1689
71. Chen BL, Liang CD, Yang J, Contreras DS, Clancy YL, Lobkovsky EB, Yaghi OM, Dai S (2006) Angew Chem Int Ed 45:1390
72. Lee JY, Pan L, Huang XY, Emge TJ, Li J (2011) Adv Funct Mater 21:993
73. Wang XF, Wang Y, Zhang YB, Xue W, Zhang JP, Chen XM (2012) Chem Commun 48:133
74. Stock N, Biswas S (2012) Chem Rev 112:933
75. Fukushima T, Horike S, Inubushi Y, Nakagawa K, Kubota Y, Takata M, Kitagawa S (2010) Angew Chem Int Ed 49:4820
76. Deng HX, Doonan CJ, Furukawa H, Ferreira RB, Towne J, Knobler CB, Wang B, Yaghi OM (2010) Science 327:846
77. Jhung SH, Lee JH, Forster PM, Ferey G, Cheetham AK, Chang JS (2006) Chem Eur J 12:7899
78. Kole GK, Vittal JJ (2013) Chem Soc Rev 42:1755
79. Lin J-B, Lin R-B, Cheng X-N, Zhang J-P, Chen X-M (2011) Chem Commun 47:9185
80. Zhu AX, Lin RB, Qi XL, Liu Y, Lin YY, Zhang JP, Chen XM (2012) Microporous Mesoporous Mat 157:42
81. Shimomura S, Matsuda R, Tsujino T, Kawamura T, Kitagawa S (2006) J Am Chem Soc 128:16416

82. Shimomura S, Horike S, Matsuda R, Kitagawa S (2007) J Am Chem Soc 129:10990
83. Chen XM, Tong ML (2007) Acc Chem Res 40:162
84. Zhang J-P, Chen X-M (2009) In: Hong M-C, Chen L (eds) Design and construction of coordination polymers, 86th edn. Wiley, New Jersey, p 63
85. Zheng YZ, Xue W, Tong ML, Chen XM, Grandjean F, Long GJ (2008) Inorg Chem 47:4077
86. Zheng YZ, Xue W, Zhang WX, Tong ML, Chen XM, Grandjean F, Long GJ, Ng SW, Panissod P, Drillon M (2009) Inorg Chem 48:2028
87. Wang YT, Fan HH, Wang HZ, Chen XM (2005) Inorg Chem 44:4148
88. Wang ZQ, Cohen SM (2009) Chem Soc Rev 38:1315
89. Cohen SM (2012) Chem Rev 112:970
90. Kitagawa S, Kitaura R, Noro S (2004) Angew Chem Int Ed 43:2334
91. Kitagawa S, Uemura K (2005) Chem Soc Rev 34:109
92. Ferey G, Serre C (2009) Chem Soc Rev 38:1380
93. Gould SL, Tranchemontagne D, Yaghi OM, Garcia-Garibay MA (2008) J Am Chem Soc 130:3246
94. He J, Yin YG, Wu T, Li D, Huang XC (2006) Chem Commun 2845
95. Zhang J-P, Horike S, Kitagawa S (2007) Angew Chem Int Ed 46:889
96. Zhang JP, Kitagawa S (2008) J Am Chem Soc 130:907
97. Xie LH, Lin JB, Liu XM, Wang Y, Zhang WX, Zhang JP, Chen XM (2010) Inorg Chem 49:1158
98. Xie LH, Wang Y, Liu XM, Lin JB, Zhang JP, Chen XM (2011) CrystEngComm 13:5849
99. Zhang JP, Zhu AX, Chen XM (2012) Chem Commun 48:11395
100. Sakata Y, Furukawa S, Kondo M, Hirai K, Horike N, Takashima Y, Uehara H, Louvain N, Meilikhov M, Tsuruoka T, Isoda S, Kosaka W, Sakata O, Kitagawa S (2013) Science 339:193
101. Murray LJ, Dincă M, Long JR (2009) Chem Soc Rev 38:1294
102. Suh MP, Park HJ, Prasad TK, Lim DW (2012) Chem Rev 112:782
103. Ma SQ, Sun DF, Simmons JM, Collier CD, Yuan DQ, Zhou HC (2008) J Am Chem Soc 130:1012
104. Bhatia SK, Myers AL (2006) Langmuir 22:1688–1700
105. Furukawa H, Miller MA, Yaghi OM (2007) J Mater Chem 17:3197
106. Matsuda R, Kitaura R, Kitagawa S, Kubota Y, Belosludov RV, Kobayashi TC, Sakamoto H, Chiba T, Takata M, Kawazoe Y, Mita Y (2005) Nature 436:238
107. Zhang J-P, Chen X-M (2009) J Am Chem Soc 131:5516
108. Li J-R, Kuppler RJ, Zhou H-C (2009) Chem Soc Rev 38:1477
109. Li JR, Sculley J, Zhou HC (2012) Chem Rev 112:869
110. Wu HH, Gong QH, Olson DH, Li J (2012) Chem Rev 112:836
111. Guo HL, Zhu GS, Hewitt IJ, Qiu SL (2009) J Am Chem Soc 131:1646
112. Venna SR, Carreon MA (2010) J Am Chem Soc 132:76
113. Sumida K, Rogow DL, Mason JA, McDonald TM, Bloch ED, Herm ZR, Bae TH, Long JR (2012) Chem Rev 112:724
114. Liao PQ, Zhou DD, Zhu AX, Jiang L, Lin RB, Zhang JP, Chen XM (2012) J Am Chem Soc 134:17380
115. Wriedt M, Sculley JP, Yakovenko AA, Ma YG, Halder GJ, Balbuena PB, Zhou HC (2012) Angew Chem Int Ed 51:9804
116. Li K, Olson DH, Seidel J, Emge TJ, Gong H, Zeng H, Li J (2009) J Am Chem Soc 131:10368
117. Chang N, Gu ZY, Yan XP (2010) J Am Chem Soc 132:13645
118. Cui XY, Gu ZY, Jiang DQ, Li Y, Wang HF, Yan XP (2009) Anal Chem 81:9771
119. Chang N, Gu ZY, Wang HF, Yan XP (2011) Anal Chem 83:7094
120. He CT, Tian JY, Liu SY, Ouyang GF, Zhang JP, Chen XM (2013) Chem Sci 4:351
121. Kurmoo M (2009) Chem Soc Rev 38:1353
122. Allendorf MD, Bauer CA, Bhakta RK, Houk RJT (2009) Chem Soc Rev 38:1330
123. Kreno LE, Leong K, Farha OK, Allendorf M, Van Duyne RP, Hupp JT (2012) Chem Rev 112:1105

124. Cheng XN, Zhang WX, Lin YY, Zheng YZ, Chen XM (2007) Adv Mater 19:1494
125. Ohba M, Yoneda K, Agusti G, Munoz MC, Gaspar AB, Real JA, Yamasaki M, Ando H, Nakao Y, Sakaki S, Kitagawa S (2009) Angew Chem Int Ed 48:4767
126. Lin JB, Xue W, Wang BY, Tao J, Zhang WX, Zhang JP, Chen XM (2012) Inorg Chem 51:9423
127. Bao X, Shepherd HJ, Salmon L, Molnar G, Tong ML, Bousseksou A (2013) Angew Chem Int Ed 52:1198
128. Lu ZZ, Zhang R, Li YZ, Guo ZJ, Zheng HG (2011) J Am Chem Soc 133:4172
129. Lu G, Hupp JT (2010) J Am Chem Soc 132:7832
130. Lu G, Farha OK, Zhang WN, Huo FW, Hupp JT (2012) Adv Mater 24:3970
131. Cui YJ, Yue YF, Qian GD, Chen BL (2012) Chem Rev 112:1126
132. Chen BL, Yang Y, Zapata F, Lin GN, Qian GD, Lobkovsky EB (2007) Adv Mater 19:1693
133. Lan AJ, Li KH, Wu HH, Olson DH, Emge TJ, Ki W, Hong MC, Li J (2009) Angew Chem Int Ed 48:2334
134. Xie ZG, Ma LQ, deKrafft KE, Jin A, Lin WB (2010) J Am Chem Soc 132:922
135. Qi XL, Lin RB, Chen Q, Lin JB, Zhang JP, Chen XM (2011) Chem Sci 2:2214
136. Ma LQ, Abney C, Lin WB (2009) Chem Soc Rev 38:1248
137. Lee J, Farha OK, Roberts J, Scheidt KA, Nguyen ST, Hupp JT (2009) Chem Soc Rev 38:1450
138. Yoon M, Srirambalaji R, Kim K (2012) Chem Rev 112:1196
139. Kawamichi T, Kodama T, Kawano M, Fujita M (2008) Angew Chem Int Ed 47:8030
140. Uemura T, Yanai N, Kitagawa S (2009) Chem Soc Rev 38:1228
141. Horcajada P, Gref R, Baati T, Allan PK, Maurin G, Couvreur P, Ferey G, Morris RE, Serre C (2012) Chem Rev 112:1232
142. Wang C, Zhang T, Lin WB (2012) Chem Rev 112:1084

Struct Bond (2014) 157: 27–88
DOI: 10.1007/430_2013_133

Published online: 7 September 2013

Luminescent Properties and Applications of Metal-Organic Frameworks

Yuanjing Cui, Banglin Chen, and Guodong Qian

Abstract Metal-organic frameworks (MOFs) are very promising multifunctional luminescent materials because of their inherent advantages of both organic linkers and inorganic metal ions, as well as the tailorability in terms of structure, dimension, size, and shape. Furthermore, the metal centers, organic linkers, metal-organic charge transfer, and guest molecules within porous MOFs all can potentially generate luminescence. Such uniqueness can allow us to generate luminescent MOF materials with systematically varied luminescence properties which are crucial for the lighting, display, and optical devices. Although luminescent MOFs are still in their infancy, the currently available results have unambiguously demonstrated that the design and construction of MOFs for luminescent functionality is very active, and hundreds of papers published in the last few years indicate that this interest is still increasing. This chapter outlines the basic principles and luminescence origins of MOF materials and then describes the latest developments in luminescent MOFs and their applications in light-emitting devices, chemical sensors, and biomedicine. As a conclusion, several evident research trends were proposed which are particularly promising but will require concerted effort for success.

Keywords Biomedicine · Chemical sensor · Luminescence · Metal-organic frameworks · White-light-emitting

---

Y. Cui and G. Qian (✉)
State Key Laboratory of Silicon Materials, Department of Materials Science and Engineering, Zhejiang University, Hangzhou 310027, China
e-mail: gdqian@zju.edu.cn

B. Chen (✉)
State Key Laboratory of Silicon Materials, Department of Materials Science and Engineering, Zhejiang University, Hangzhou 310027, China

Department of Chemistry, University of Texas at San Antonio, San Antonio, TX 78249-0698, USA
e-mail: Banglin.Chen@utsa.edu

Contents

Abbreviations

1,2-BDC	1,2-Benzenedicarboxylate
1,3-BDC	1,3-Benzenedicarboxylate
1,4-BDC	1,4-Benzenedicarboxylate
1,4-NDC	1,4-Naphthalenedicarboxylate
2,3-pydcH_2	Pyridine-2,3-dicarboxylic acid
2,4-DNT	2,4-Dinitrotoluene
2,6-pydc	Pyridine-2,6-dicarboxylate
3-Abpt	4-Amine-3,5-bis(3-pyridyl)-1,2,4-triazole
3-PyHBIm	2-(3-Pyridyl)benzimidazole
4-Ptz	5-(4-Pyridyl)tetrazole
4-PyHBIm	2-(4-Pyridyl)benzimidazole
5MT	5-Methyl-tetrazole
ABTC	3,3′,5,5′-Azobenzenetetracarboxylate
ATA	5-Amino-tetrazole
ATPA	2-Aminoterephthalate
BCPA	9,10-Bis(*p*-carboxyphenyl) anthracene
BDC-F_4	2,3,5,6-Tetrafluoro-1,4-benzenedicarboxylate
bhc	Benzenehexacarboxylate
bipy	4,4′-Bipyridine
bpdc	4,4′-Biphenyldicarboxylate
Bpe	*Trans*-bis(4-pyridyl)ethylene
bpee	1,2-Bipyridylethene
bpp	1,3-Bis(4-pyridyl)propane
BPT	Biphenyl-3, 4′, 5-tricarboxylate
bpydc	2,2′-Bipyridine-5,5′-dicarboxylate

bpz	3,3′,5,5′-Tetramethyl-4,4′-bipyrazolate
bpze	Bis(pyrazole-1-yl)ethane
bpzm	Bis(pyrazole-1-yl)methane
bpzp	Bis(pyrazole-1-yl)prothane
BTAH	Benzotriazole
BTB	4,4′,4″-Benzenetribenzoate
BTC	1,3,5-Benzenetricarboxylate
CCT	Correlated color temperature
chdc	1,4-Cyclohexanedicarboxylate
CIE	Commission International ed'Eclairage
CRI	Color rendering index
CT	Computed tomography
CTAB	Cetyltrimethylammonium bromide
CTC	*Cis,cis*-1,3,5-cyclohexanetricarboxylate
DABCO	1,4-Diazabicyclo[2.2.2]octane
dbm	Dibenzoylmethane
DFDA	9,9-Dipropylfluorene-2,7-dicarboxylate
DHT	2,5-Dihydroxyterephthalate
dipicH_2	Dipicolinic acid
DMA	Dimethylammonium
DMBDC	2,5-Dimethoxy-1,4-benzenedicarboxylate
DMNB	2,3-Dimethyl-2,3-dinitrobutane
dpbp	4,4′-Bis(diphenylphosphoryl) biphenyl
dpNDI	*N,N*′-di(4-pyridyl)-1,4,5,8-naphthalenediimide
ESCP	Ethoxysuccinato-cisplatin
FMA	Fumarate
H_2dtba	2,2′-Dithiobisbenzoic acid
H_2ox	Oxalic acid
H_2pda	1,4-Phenylenediacetic acid
H_2PhenDCA	1,10-Phenanthroline-2,9-dicarboxylic acid
H_2-PVDC	4,4′-[(2,5-Dimethoxy-1,4-phenylene)di-2,1-ethenediyl]bis-benzoic acid
H_3bidc	1H-benzimidazole-5,6-dicarboxylic acid
H_3NTB	4,4′,4″-Nitrilotrisbenzoic acid
hfa	Hexafluoro acetylacetonato
HMOFs	Heterometal-organic frameworks
Hnic	Nicotinic acid
IDC	Imidazole-4,5-dicarboxylate
ILAG	Ion- and liquid-assisted grinding
imdc	4,5-Imidazoledicarboxylate
LAG	Liquid-assisted grinding
LMCT	Ligand-to-metal charge transfer
Ln^{3+}	Lanthanide ions
MES	2-(*N*-morpholino)ethanesulfonate

MLCT	Metal-to-ligand charge transfer
MOFs	Metal-organic frameworks
MRI	Magnetic resonance imaging
nds	2,6-Naphthanlenedisulfonate
NIR	Near-infrared
NMOFs	Nano-MOFs
NMP	1-Methyl-2-pyrrolidone
NTA	Nitrilotriacetate
OBA	4,4′-Oxybis(benzoate)
ox	Oxalate
PDA	Pyridine-2,6-dicarboxylic acid
pdc	2,5-Pyridinedicarboxylate
Phen	1,10-Phenanthroline
$PhPPy_2$	Bis(2-pyridyl)phenylphosphine
pimda	2-Propyl-4,5-imidazole-dicarboxylate
Pmtz	5-(Pyrimidyl)-tetrazolato
ppy	2-Phenylpyridine
pta	2,4,6-Pyridinetricarboxylate
pydc	Pyridine-dicarboxylate
Pz	3,5-Dimethylpyrazole
pzdc	2,5-Pyrazinedicarboxylate
QD	Quantum dot
quin	2,2′-Biquinoline
Rh6G	Rhodamine 6G
SBUs	Secondary building units
sfdb	4,4′-Sulfonyldibenzoic acid
TAA	1-H-tetrazole-5-acetate
TBAPy	1,3,6,8-Tetrakis(*p*-benzoic acid)pyrene
TDC	Thiophene-2,5-dicarboxylate
TEAF	Triethylammonium fluoride
TNT	2,4,6-Trinitrotoluene
tta	2-Thenoyltrifluoroacetone
TzC	5-Carboxylato-tetrazolato
UV	Ultraviolet
VOCs	Volatile organic compounds

1 Introduction

Metal-organic frameworks (MOFs), also known as coordination polymers or coordination networks, have attracted tremendous attention over the last decade due to their promising applications in gas storage [1, 2], gas capture [3], separations [4], chemical catalysis [5], luminescence [6–8], magnetism [9], and drug delivery [10]. MOFs are highly crystalline inorganic–organic hybrid materials comprised of metal ions or clusters that form vertices of a framework and organic linkers that act as bridges. In most cases, MOFs are rigid enough to allow the removal of solvent or other guest molecules in the pores resulting in permanent porosity. Thus, these materials generally exhibit microporous character, and the pore sizes could be tuned from several angstroms to several nanometers by altering the length of organic linkers or controlling the interpenetration. The extraordinarily high porosity, very large surface areas, and wide range of pore sizes of MOFs make them attractive for scientific and industrial application as new porous materials [11–15]. Compared with conventional porous materials, such as zeolites, carbon-based materials, one main advantage of MOFs is their easily modifiable synthesis to control the size, shape, dimension, connectivity, and surface of the pores by changing the organic ligands and metal ions. A variety of functionality can be incorporated in the MOFs by varying the coordination geometries of the metals and the topicities of the ligands, modifying the pore surface, or introducing functional groups. These features are attractive enough to attract more research effort toward the development of MOFs with designable framework structures and tunable functionality for specific applications.

Although some early reports on MOFs have appeared in 1960s, the interest in this field was kindled when Yaghi et al. reported the structure of the well-known MOF-5 with significantly high surface area of 2,900 m^2/g in 1999 [16]. The implementation of simple solvothermal synthesis and incorporation of large organic linkers to assemble highly porous MOF-14 in 2001 [17], to some extent, has highlighted the establishment of a general strategy to construct porous MOFs. Since then, much research effort has been devoted for the investigation of the high porosity, stability, control of pore shape and size, and framework flexibility. Up to date, thousands of different MOFs have been synthesized, and the number of publications related with MOFs has increased remarkably.

Recently, the development of luminescent functional MOFs has attracted more scientific interest because both the inorganic and organic moieties can provide the platforms to generate luminescence, while metal–ligand charge-transfer-related luminescence within MOFs can add another dimensional luminescent functionality. Furthermore, some guest molecules within MOFs can also emit and/or induce luminescence [7, 18–20]. The infinite three-dimensional (3D) framework structures, inherent microporosity, and high surface areas offer great opportunities for intermolecular interaction between MOFs and guest molecules, which can dramatically vary the coordination environment and energy transfer, thus modulating the luminescent properties of MOFs. Moreover, the pore surface and framework structure of the MOFs can be further functionalized by postsynthetic modification, which enables the energy level and charge transfer to be finely tuned targeting a specific luminescent property. Compared with the traditional molecular approach, by making use of molecular organic and coordination compounds to explore functional luminescent materials, the MOF approach also has the advantage to collaboratively interplay/interact with each other among the periodically located luminescent centers to direct new functionalities. These advantages lead to the vast versatility of luminescent MOFs in practical applications, which has exceeded the capabilities of many other luminescent materials.

During the past 10 years, there is an increasing trend on the exploration and discovery of functional luminescent MOFs. Studies on luminescent MOFs up to now mainly include the following: investigating the fundamental synthesis and luminescent properties of MOFs; creating MOFs with tunable luminescence for application in light-emitting and display devices; using luminescence of MOFs to probe local environment, structure, and guest species; and developing multifunctional MOFs combining luminescence, magnetism, and biocompatibility for biomedical application. The permanent porosity with some luminescent MOFs has enabled their reversible storage and release of guest substrates and provided the hosts for their differential recognitions with sensing species, thus distinguishing porous luminescent MOFs from the traditional inorganic and organic luminescent materials in terms of their multifunctionalities. The MOF approach can also offer a variety of other attractive characteristics such as the straightforward syntheses, predictable structures and porosities, nanoscale processibility, and collaborative properties to develop luminescent functional MOF materials. The chapter describes the latest developments in luminescent MOF materials and their applications in light-emitting devices, chemical sensors, and biomedicine. As a conclusion, several evident research trends were proposed which are particularly promising but will require concerted effort for success. The purpose of this chapter is to give a comprehensive overview of the different types of luminescent MOFs, particularly those with interesting and important applications. The definition of MOFs is interpreted rather broadly in this review, so some nonporous coordination polymers will be also included.

2 Construction and Modification of MOFs

2.1 *Syntheses of MOFs*

MOFs are essentially constructed through a simple approach known as the building-block methodology, which relies on the assembly of molecular or ionic moieties into extended arrays using complementary interactions between components. In the construction of MOFs, one-dimensional (1D) chain structure can be achieved through the connecting of metal ions that prefer linear coordination with simple organic linkers, while two-dimensional (2D) sheet structures and 3D architectures can be built using trigonal or square-planar metal center nodes and tetrahedral or octahedral metal centers with the same organic linkers, respectively. Thus, by changing the metal geometry and the organic linker type, a wide variety of different MOFs can be generated. In addition, the in situ formation of some metal-containing clusters (generally termed as secondary building units (SBUs)) with predetermined coordination geometries during the MOF syntheses has enabled us to synthesize a series of MOFs with predictable structures. For example, the connectivity of $Zn_4O(COO)_6$ SBUs with a large number of bicarboxylate led to the so-called isoreticular MOFs (IRMOFs) with the default cubic structures [21]. Another well-known example of isoreticular MOFs are those assembled from paddle-wheel $M_2(COO)_4$

(M=Cu^{2+}, Zn^{2+}, Co^{2+}, and Ni^{2+}) with bicarboxylate ($R(COO)_2$) and pillar organic linker (L) to form distorted primitive cubic nets of $M_2(ROO)_2L$ [22, 23].

The compound 4,4′-bipyridine or its derivatives and multidentate carboxylic acids with suitable spacers, especially benzene-multicarboxylic acids, are the prominent linkers for constructing diverse architectures. The 4,4′-bipyridine-like ligands are simple linear linkers which possess monodentate donor nitrogen atoms at opposite ends of the molecule; therefore the determination of framework is directly governed by the coordination environment and geometry of the metal ion. Multidentate carboxylic acids are also attractive linkers because their high acidity allows facile in situ deprotonation. In addition, the ability of carboxylic acids to form more than one coordination bond increased the chances of generating a robust porous framework that retains the MOF integrity in the absence of guest molecules. Yaghi et al. have designed and synthesized sixteen MOFs, denoted as IRMOF-*n* (n = 1–16), with wide ranges of pore size and chemical functionality through the reaction of various benzenedicarboxylate ligands with Zn^{2+} ions. The framework consists of oxo-centered Zn_4O tetrahedron cluster located at the edges of the cube and bridged by six carboxylate units to give secondary building units (SBUs), which are further connected by the bridging dicarboxylate, resulting in the 3D cubic networks. The MOFs with bigger cavities or channels can be synthesized by using longer benzenedicarboxylate linkers [21]. Besides the rigid linkers as mentioned above, flexible ligands such as cyclohexane-1,2,5,6-tetracarboxylic acid, tetrakis(imidazole), and *N*,*N*′-diacetic acid imidazolium can be used in the synthesis of MOFs [24–27]. These linkers are prone to give nonporous structures through dense packing of a single network or through interpenetration of two or more networks.

Generally, MOFs can be produced by mixing two solutions containing the metal ions and the organic ligands, either at room temperature or under hydrothermal/solvothermal conditions [28–30]. Initial syntheses of coordination polymers/metal-organic frameworks were realized by the diffusion technique in which the solution containing metal salts was slowly diffused into the solution containing organic linkers such as 4,4-bipyridyl, and/or the weak bases such as triethylamine were slowly diffused into the solution containing metal salt and bridging organic carboxylic acid [16]. Such synthesis approach is very time-consuming while the yield is typically low. In fact, the yield to synthesize MOF-5 in the first published paper is only about 5% [16]. Hydrothermal syntheses of MOFs have been feasible to those reaction systems in which organic linkers can be partially soluble in water at higher temperature. The implementation of solvothermal syntheses by making use of organic solvents to readily dissolve organic linkers, as exemplified in the synthesis of MOF-14 in 2001, has facilitated the syntheses of crystalline MOF materials significantly [17]. Generally speaking, heating the mixture containing inorganic salts and organic linkers in certain types of solvents (DMF is the most popular solvent) at the certain temperature (60–120°C) for several hours to two days can readily lead to crystalline MOFs for structure characterization and property exploration. The simple and straightforward syntheses of MOFs, to some of the extent, have allowed many chemists and materials scientists readily to start their research programs in such a very active research field [11, 20, 31–33]. The solvothermal synthesis approach has also enabled the exploration of combinatorial approach to screen and discover new MOF materials within a very short period of time, as exemplified in the combinatorial syntheses of a series of zeolitic imidazole frameworks (ZIFs, a special subclass of MOFs) [34–39]. BASF has even been able to produce and commercialize some prototypical MOFs in large scale.

In recent years, other alternative synthetic methods including mechanochemical methods of solid-state grinding and liquid-assisted grinding, sonochemical methods, and microwave-assisted synthesis have also been emerged. In mechanochemical synthesis, a mixture of organic linkers and metal salts is ground together in a mechanical ball mill to yield the MOFs [40–49]. Compared with solvothermal synthesis, the obvious advantage of mechanochemical synthesis is the ability to avoid the use of solvent. In addition, sufficient amounts of pure MOFs for broad-range testing can be easily produced. Klimakow et al. demonstrated the mechanochemical synthesis of the intensively studied MOFs HKUST-1 ($Cu_3(BTC)_2$, BTC = 1,3,5-benzenetricarboxylate) and MOF-14 ($Cu_3(BTB)_2$, BTB = 4,4′,4″-benzenetribenzoate) [41]. The obtained MOFs exhibit the high surface area which is comparable to the highest given values in the literature for the respective compounds. Friščić et al. introduced an improved mechanochemical approach, denoted as ion- and liquid-assisted grinding (ILAG) [44]. The construction of MOFs is accelerated and directed by small amounts of salts, for example, nitrate and sulfates induce the formation of tetragonal and hexagonal structure, respectively.

Sonochemistry was recently explored by Son et al. in the synthesis of the popular MOF-5 from solutions of zinc nitrate and terephthalic acid [50]. Compared with conventional solution synthesis, sonochemistry method can lead to homogeneous nucleation and a significant reduction in crystallization time [51–54]. By using 1-methyl-2-pyrrolidone (NMP) as solvent, high-quality MOF-5 crystals of 5–25 μm in size are produced, with similar physicochemical properties to those crystals prepared under conventional convective heating. Similarly, high-quality MOF-177 crystals, namely, $Zn_4O(BTB)_2$ (BTB = 4,4′,4″-benzenetribenzoate), are also synthesized via a sonochemical route in the presence of NMP as a solvent [55]. The synthesis time is shortened drastically from 48 h to ca. 0.5 h accompanied by a substantial reduction in crystal size to 5–20 μm. The product yield of MOF-177 synthesized via the sonochemical route was 95.6%, which was significantly higher than the product yields of other methods.

Microwave-assisted solvothermal synthesis involves heating a solution with microwaves to produce nanosized crystals, which allows high-quality MOF crystals to be synthesized in under a minute [56–65]. Although the microwave method usually cannot produce single crystals with a big size for structural analysis, its homogeneous effects could create a uniform seeding condition; thus the size and shape of the crystals can be well controlled and the rate of chemical reactions can be accelerated. The application of microwave radiation to conduct MOF synthesis has been investigated by Ni et al. in the construction of three known MOFs, namely, IRMOF1 (MOF-5), IRMOF2, and IRMOF3 [65]. The resulting MOF crystals exhibit identical cubic morphology and a very uniform distribution of particle sizes, which can be varied from micrometer down to submicrometer scale by controlling the reactant concentration. The uniform size and shape of MOFs are attributed to the ability of the microwave technique to control the nucleation process throughout the volume of solution, in which all of the crystals are nucleated at once. Recently, the use of microwave-assisted synthesis to construct MIL-47 and six new vanadium MOFs has also been reported by Centrone et al [66].

2.2 *Nanoscale MOFs*

Although MOFs have shown promise for a number of practical applications, the bulk crystalline nature do not always fulfill all specific needs for these applications. For example, in the fields of biology, drug delivery, and biomedical imaging, the nanoscale dimensions are required to provide materials of sufficiently small sizes for their internalization into cells [67]. Recently, scaling down these framework materials has afforded an exciting new class of materials known as nano-MOFs (NMOFs) [68–71], which exhibit the combination of the rich diversity of compositions, structures, and properties of bulk MOFs with the obvious advantages of nanomaterials. The limited solution-based behavior of bulk MOFs can be overcome by the preparation of nano-MOFs, enabling the MOFs highly dispersed in aqueous media or other solvents and coated for improving their biocompatibility. In addition, the NMOFs may display the higher surface areas than their macroscopic counterparts and the unique size-dependent optical, electrical, and magnetic properties.

Usually, NMOFs can be produced through two major strategies [67]. One is the controlled precipitation of self-assembled MOFs once they are formed, which includes the use of microwaves or sonochemistry methods. As mentioned above, the microwave radiation and ultrasounds can accelerate the nucleation and increase the seed number, avoiding the fast growth of MOF crystals. Therefore, microwave and sonochemistry synthesis methods are the productive strategies generally used to produce a wide variety of NMOFs. The other method is confining the assembly process at specific nanoscale locations. For example, the supramolecular assembly of NMOFs can be confined into droplets by using microemulsion method or controlled on the surface of substrates by using template. Microemulsion is a suspension consisting of small droplets of an immiscible phase (nonpolar or polar) dispersed in a continuous phase, usually stabilized by a surfactant. Microemulsion system can be divided into oil-in-water and water-in-oil, depending on the concentration of the different components. By varying the concentration of the dispersed phase and the surfactant, the size of the droplets can be tuned in the range 50–1,000 nm, approximately. These nanodroplets in the microemulsion can be used as "nanoreactors" to control the kinetics of particle nucleation and growth [72].

The utilization of water-in-oil, or reverse, microemulsion method for NMOFs synthesis has been reported by Rieter et al. [73], who prepared a NMOF $Gd(BDC)_{1.5}(H_2O)_2$ (BDC = 1,4-benzenedicarboxylate) by reacting $GdCl_3$ and bis(methylammonium)benzene-1,4-dicarboxylate in the cationic cetyltrimethylammonium bromide (CTAB)/

isooctane/1-hexanol/water system. By varying the w value (defined as the water/surfactant molar ratio) of the microemulsion systems, the morphologies and sizes of the NMOFs can be tuned. Furthermore, the particle size of NMOF is also affected by the reactant concentration and the reactant ratio. The crystalline nanoplates of [Gd(1,2,4-BTC)$(H_2O)_3$]·H_2O (BTC = benzenetricarboxylate), with a diameter of ~100 nm and an average thickness of 35 nm, were also prepared using the reverse microemulsion method. This group also synthesized the NMOFs Mn(BDC)$(H_2O)_2$ and $Mn_3(BTC)_2(H_2O)_6$ in reverse microemulsions [74]. The reaction of $MnCl_2$ and $[NMeH_3]_2$(BDC) in a CTAB/1-hexanol/*n*-heptane/water microemulsion with a *w* value of 5 yielded the nanorods of Mn(BDC)$(H_2O)_2$, with a structure corresponding to a known crystalline phase. NMOF of $Mn_3(BTC)_2(H_2O)_6$ can be produced in a CTAB/1-hexanol/isooctane/water microemulsion ($w = 10$) with a Na_3(BTC)/$MnCl_2$ molar ratio of 2:3. The particles of $Mn_3(BTC)_2(H_2O)_6$ are crystalline spiral rods, but do not correspond to any known phase of Mn-BTC-based MOFs. The similar synthesis of luminescent NMOF $Eu_2(BDC)_3(H_2O)2{\cdot}(H_2O)_2$ by reacting $EuCl_3$ and the methylammoium salt of H_2BDC in a CTAB/1-hexanol/isooctane/water microemulsion was reported by Xu et al. [75].

Besides the room-temperature microemulsion synthesis, the surfactant-assisted synthesis at elevated temperatures has been developed by the Lin group for the synthesis of NMOFs [76]. The blocklike nanoparticles of Gd_2(bhc)$(H_2O)_6$ (bhc = benzenehexacarboxylate) were synthesized by heating a CTAB/1-hexanol/*n*-heptane/water microemulsions containing $[NMeH_3]_6$(bhc) and $GdCl_3$ at 120°C for 18 h. Interestingly, the reaction of $GdCl_3$ and mellitic acid in the same microemulsion system at 60°C in a microwave reactor for 15 min afforded the high-aspect ratio nanorods, which has a formula of Gd_2(bhc)$(H_2O)_8{\cdot}2(H_2O)$. Additionally, the surfactant-assisted synthesis under microwave heating is also utilized to produce the Mn-based NMOFs, which have been synthesized in reverse microemulsions at room temperature, as mentioned above. Nanorods of Mn(BDC)$(H_2O)_2$ synthesized under microwave heating have identical phase and similar morphology to those obtained from the microemulsion, while the particles of $Mn_3(BTC)_2(H_2O)_6$ obtained under microwave heating have a blocklike morphology, with lengths ranging from ~50 to 300 nm in the three dimensions [74].

2.3 Postsynthesis Modification

The decoration of the pores and modification of the structures afford the possibility to modulate the physical and chemical properties of MOFs and tune the interactions of guest species with host frameworks, enabling the MOFs useful for more specialized and sophisticated applications. Modification and functionalization of MOFs have been reported by using unsaturated metal centers or functional groups on the organic bridging ligands [77–81]. Although the method of using modified ligand directly in synthesis has resulted in MOFs with pendant groups such as −Br, $-NH_2$, and $-CH_3$ decorating the pores, it is limited because some kinds of functional groups, such as −OH, −COOH, and N-donating groups, have a tendency to coordinate to metal ions. In addition, the incorporation of functional groups on the linking ligands may lead to insolubility, thermal instability, or steric bulk, hindering the formation of desirable MOF structures. Postsynthesis modification has recently emerged as a highly versatile tool to functionalize MOFs, which allows introducing covalent attachment to the organic linker after the framework has been synthesized. The realization of postsynthesis approach within MOFs has certainly enriched the MOFs with predictable structures [82]. Incorporation of functional organic and/or metal-organic sites within MOFs is not only very important to tune the porosity and introduce their specific recognition of substrates, but also crucial to systematically modulate their luminescent functionalities.

The MOF IRMOF-3, which is constructed from Zn_4O SBUs and 2-amino-1,4-benzenedicarboxylic acid (NH_2–BDC) linkers, has been used to demonstrate the covalent postsynthesis modification by Cohen et al. [83, 84]. This MOF can be reacted with alkyl anhydrides to yield a series of amide-bearing MOFs with different substituents, but maintain the same crystallinity and thermal stability of IRMOF-3. This group also incorporated a series of medium to long alkyl groups within IRMOF-3 through postsynthesis method to yield the hydrophobic MOFs [85]. In the same study, the MOF MIL-53(Al)–NH_2 was also modified with three different alkyl anhydrides to generate MIL-53(Al)–AM1, MIL-53(Al)–AM4, and MIL-53 (Al)–AM6. The MOF MIL-53(Al)–NH_2 and MIL-53(Al)–AM1 are hydrophilic with contact angles of 0°, while MIL-53(Al)–AM4 and MIL-53(Al)–AM6 show superhydrophobic properties with contact angles greater than 150°. This was proposed that the combination of the submicrometer crystallite size and the hydrophobic functional groups result in the superhydrophobicity. The similar modification was also executed using the MOFs IRMOF-3, DMOF-1–NH_2, and UMCM-1–NH_2 to enhance hydrogen sorption properties of MOFs [86]. Taylor–Pashow et al. have also utilized the covalent postsynthesis method to attach drugs and imaging contrast agents to a Fe-based MOF, enabling it suitable for imaging and therapeutic applications [87].

The utilization of "click" chemistry for postsynthesis modification of MOFs has been reported by Goto et al. [88], who reacted a MOF bearing the azide group in the organic linkers with some alkynes containing various functional groups through a copper(I)-catalyzed azide–alkyne cycloaddition reaction. Savonnet et al. developed a one-pot, two-step modification method starting from already available amino-derived MOFs, DMOF–NH_2 [Zn(bdc–NH_2)(DABCO)] and MIL-68(In)–NH_2 [In (OH)(bdc–NH_2)] [89]. In this method, the amino groups of DMOF–NH_2 were transformed into an azide DMOF–N_3 and then reacted with an excess of phenylacetylene and $Cu(CH_3CN)_4PF_6$ via "click" reaction to generate the triazole-modified DMOF. The modification was also executed for the MIL-68(In)–NH_2, and the same results was obtained. A method to selectively modify both the surface and interior of a MOF has been developed by Gadzikwa et al. [90], who prepared a non-catenated, pillared paddle-wheel framework TO-MOF using 1,2,4,5-tetrakis (4-carboxyphenyl)benzene (TCPB), 3-[(trimethylsilyl)ethynyl]-4-[2-(4-pyridinyl) ethenyl]pyridine, and Zn^{2+} in acidic DMF at 80 °C for 1day. The trimethylsilyl (TMS) groups within To-MOF can be easily removed via deprotection, and the exposed terminal alkyne groups can react with benzyl azide to form a triazole via the "click" reaction. In order to perform a selective modification, To-MOF was firstly exchanged with $CHCl_3$ and then exposed in an aqueous solution of KF to yield the surface-deprotected MOF, because the KF cannot enter the pores which are filled with $CHCl_3$. Thus, only the deprotected surface of MOF can be reacted with the dye ethidium bromide monoazide to obtain the surface-modified MOF. Subsequently, the interior of the MOF can be deprotected using triethylammonium fluoride (TEAF) and further reacted with benzyl azide to achieve the differentially functionalized MOF.

3 Luminescent Properties of MOFs

So far, hundreds of luminescent MOFs have been reported, and the luminescence can exist in several forms. Luminescence can arise from direct organic ligands excitation (particularly from the highly conjugated ligands), metal-centered emission (widely observed in lanthanide MOFs through the so-called antenna effect), and charge transfer such as ligand-to-metal charge transfer (LMCT) and metal-to-ligand charge transfer (MLCT). Furthermore, the guest molecules can also result in luminescence onto MOFs.

3.1 Ligand-Based Luminescence

Generally, when an organic molecule absorbs a photon of appropriate energy, a chain of photophysical events occur, including internal conversion or vibrational relaxation, fluorescence, intersystem crossing, and phosphorescence. These processes can be divided into radiative and nonradiative processes. Nonradiative processes include internal conversion, vibrational relaxation, and intersystem crossing. Internal conversion is defined as the nonradiative transition between different electronic states of the same multiplicity, whereas intersystem crossing is defined as that between states of different multiplicity. Vibrational relaxation is a process in which the content of the vibrational energy in a given electronic state changes. This process includes vibrational-energy deactivation and vibrational-energy redistribution. The former process accompanies energy loss of the excited molecules, whereas the latter is normally an isoenergetic process in which excess energy in given vibrational modes redistributes to other modes. Radiative processes of organic molecules include fluorescence and phosphorescence emission. Fluorescence is defined as a radiative transition between electronic states of the same multiplicity, such as the radiative transition from its first singlet state S_1 to the ground singlet state S_0, which is spin allowed. Phosphorescence of organic molecules corresponds to the radiative transition from triplet state T_1 to S_0, which is spin forbidden and has lifetime on the order of several microseconds to seconds.

The luminescence properties can be characterized by the following parameters: (1) Luminescence spectra, defined as fluorescence intensity as a function of a wavelength; (2) Quantum yield, the ratio of the total number of emitted photons released in the process of fluorescence to the total number of molecules promoted to the excited state. The emission quantum yield of all emissive organic molecules is <1. This observation implies that there are nonradiative processes other than radiative processes. Further, it was revealed that nonradiative processes are essentially intramolecular in nature even under the collision-free conditions; and (3) Lifetime, which refers to the average time the molecule stays in its excited state before emitting a photon and is determined as being inversely proportional to the sum rate constants of a radiative process and the nonradiative processes collectively known as quenching [91].

The ligands in MOFs are commonly π-conjugated organic molecules due to their rigidity, and the majority of them are based on rigid backbones functionalized with multi-carboxylate groups or heterocyclic groups for metal–ligand coordination. Usually, the fluorescence emission of organic ligands is similar to their emission behavior in solution, corresponding to the transition from the lowest excited singlet state to the singlet ground state, and the transitions are either $\pi \rightarrow \pi^*$ or $n \rightarrow \pi^*$ in nature. However, the fluorescence properties such as maximum emission wavelength and lifetime of organic linkers in solid MOFs are often different from those of the free molecules. This is because the organic linkers are stabilized within MOFs, which reduces the nonradiative decay rate and leads to increased fluorescence intensity, lifetimes, and quantum efficiencies. In the solid state, molecular interactions make the molecules close together, which enable charge transfer among the organic ligands/linkers, resulting in shift of spectra, broadening of the emission, and loss of fine structure. In addition, the size and nature of metal ions, the orientation and arrangement of linkers, and the coordination environment within MOF can affect the fluorescence properties of the organic linkers because these factors will induce their different intramolecular or intermolecular interactions among organic linkers. Therefore, controlling these interactions is crucial to tune the luminescence properties of MOFs for a particular application. The following examples show how intramolecular/intermolecular interactions of organic linkers have played the important roles on the luminescence properties of the resulting MOFs.

The MOF $Zn_3(\mu_5\text{-pta})_2(\mu_2\text{-}H_2O)_2$ (pta = 2,4,6-pyridinetricarboxylate) exhibits strong luminescence at 467 nm [92], while the free ligand molecule displays a weak luminescence at 415 nm when excited at 338 nm at room temperature. The fluorescent enhancement and red shift of organic linkers within this MOF is attributed to the formation of the framework structure which has enabled the rigidity of the aromatic backbones and maximized the intramolecular/intermolecular interactions among the organic linkers for their energy transfer and decreased the intraligand HOMO–LUMO energy gap. Coordination-perturbed ligand-centered luminescence was reported in MOFs $Zn_3(BTC)_2(DMF)_3(H_2O)\cdot(DMF)(H_2O)$ and $Cd_4(BTC)_3(DMF)_2(H_2O)_2\cdot 6H_2O$ (BTC = 1,3,5-benzenetricarboxylate) [93]. The strongest emission at 370 nm of the free BTC when excited at 334 nm, attributed to the $\pi^* \rightarrow n$ transitions, was shifted to 410 and 405 nm in these two MOFs, respectively, when excited at 341 and 319 nm.

Usually, most organic molecules are highly emissive in solution but become weak or even totally annihilated when the molecules are aggregated due to energy transfer and the formation of excimers and exciplexes. However, some molecules can exhibit the opposite effect: they show no emission when dissolved in good solvents but become highly emissive upon aggregation in poor solvents or in the solid state. This process is termed as "aggregation-induced emission (AIE)," which makes it possible to actively utilize the aggregation process, instead of passively working against it. Recently, Dinca et al. describe the new coordination-induced emission effect (MCIE) through coordinative immobilization of tetrakis (4-carboxyphenyl)ethylene (TCPE) within rigid porous MOFs [94]. Although with the absence of close-packed tetraphenylethylene cores, which are required for fluorescence "turn-on" in AIE, the MOFs $Zn_2(TCPE)(H_2O)_2{\cdot}4DEF$ and $Cd_2(TCPE)\ (DEF)(C_2H_5OH)_2{\cdot}DEF$ exhibit strong emission at 480 and 455 nm, respectively, similar to the value observed for solid H_4TCPE, whose emission maximum is 480 nm (Fig. 1). Furthermore, both the MOFs exhibit biexponential fluorescence decays with values similar to those observed for solid H_4TCPE. This result demonstrates that anchoring AIE-type molecules to metal ions within a MOF is also an effective method to turn-on fluorescence by restricting the rotation of the phenyl rings.

3.2 Lanthanide Luminescence

Lanthanide ions (Ln^{3+}) are characterized by a gradual filling of the 4f orbitals, from $4f^0$ (for La^{3+}) to $4f^{14}$ (for Lu^{3+}). The valence 4*f* electrons of trivalent lanthanide ions, Ln(III), are well shielded from the environment by the outer core 5*s* and 5*p* electrons and are minimally involved in bonding [95]. Because of this shielding, the atomic properties of these ions are typically retained after complexation. The absorption and emission spectra of Ln(III) ions consist of sharp, narrow bands corresponding to the f–f transitions of the metal ion. Figure 2 shows the most significant energy levels for several lanthanide ions. The f–f transitions are Laporte forbidden, resulting in low molar absorption coefficients in the near-UV and visible ranges ($<10\ M^{-1}\ cm^{-1}$) and long luminescence lifetimes in the millisecond range. The number of observed bands depends on the particular lanthanide ion and its arrangement of electronic states. Except for lanthanum (La^{3+}) and lutetium (Lu^{3+}), each lanthanide ion has its specific emission bands: for example, the lanthanide ion europium (Eu^{3+}) emits red light; terbium (Tb^{3+}) green light; and neodymium (Nd^{3+}), ytterbium (Yb^{3+}), and erbium (Er^{3+}) near-infrared light.

The lanthanide ions suffer from weak light absorption due to the forbidden f–f transitions, making the direct excitation of the metals very inefficient unless high-power laser excitation is utilized. This problem can be overcome by coupling species that can participate in energy transfer processes, known as "luminescence sensitization" or "antenna effect" [96–98]. The mechanism of antenna sensitization within MOFs is comprised of three steps: light is absorbed by the organic ligands around the lanthanide ions; energy is transferred to the lanthanide ions from organic

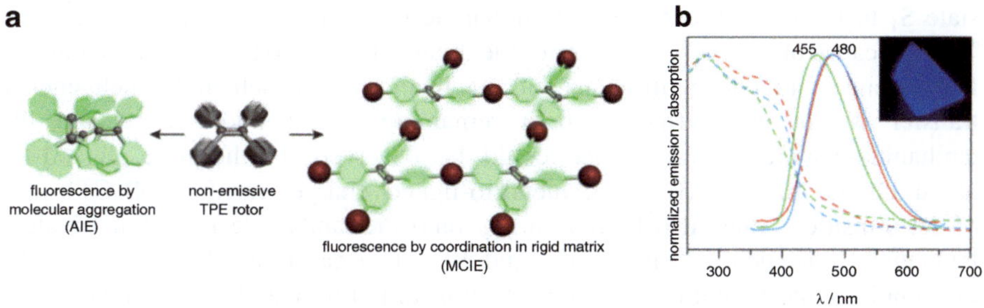

Fig. 1 (**a**) Turn-on fluorescence in a TPE rotor by aggregation (AIE) and by coordination in a rigid MOF matrix (MCIE) and (**b**) Diffuse reflectance and emission spectra of H_4TCPE (*blue*), $Zn_2(TCPE)(H_2O)_2{\cdot}4DEF$ (*red*), and $Cd_2(TCPE)(DEF)(C_2H_5OH)_2{\cdot}DEF$ (*green*). Reprinted with the permission from [94], Copyright 2011, American Chemical Society

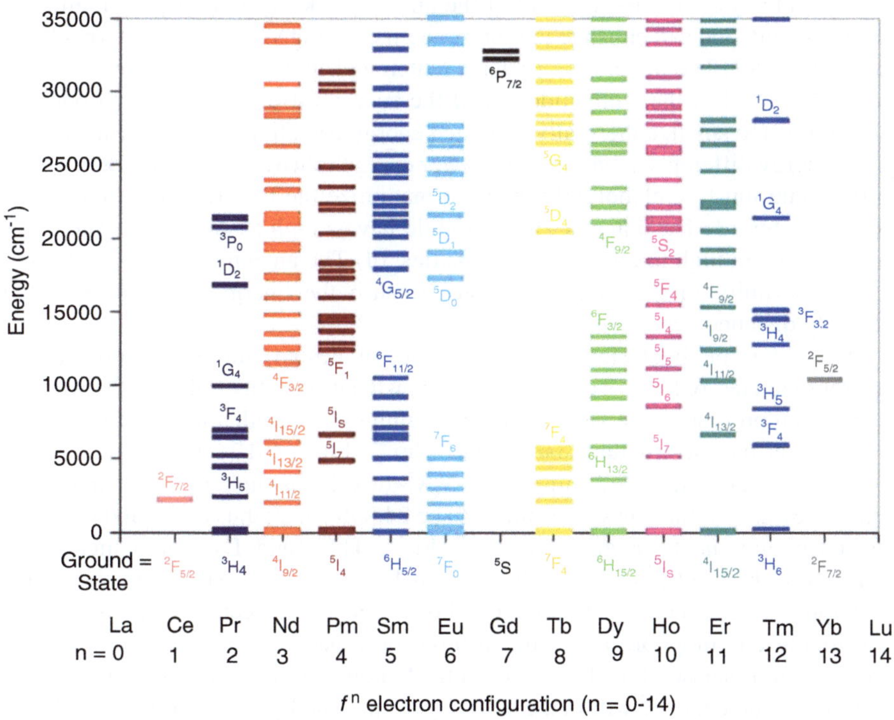

Fig. 2 A summary of electronic excited-state energy levels for Ln^{3+} ions. Reprinted with the permission from [95], Copyright 2009, American Chemical Society

ligands; and then luminescence is generated from the lanthanide ions. One of the main energy migration paths is ligand-centered absorptions followed by intersystem crossing $S_1 \rightarrow T_1$, $T_1 \rightarrow Ln^{3+}$ transfer, and metal-centered emission. Another possible path is the direct transfer of energy from the excited singlet state S_1 to the energy levels of the lanthanide ion, which is known for Eu^{3+} and Tb^{3+}. Because of the energy transfer of the organic ligands to the lanthanide ion, the fluorescence and phosphorescence of the ligands is quenched. If such energy transfer is not very efficient, both remaining ligand fluorescence and the lanthanide-centered luminescence could be observed. Furthermore, ligand-to-metal charge transfer (LMCT), metal-to-ligand charge transfer (MLCT), and 4f–5d transitions may also funnel energy onto the lanthanide ions. For trivalent lanthanide ions such as Sm^{3+}, Eu^{3+}, and Yb^{3+} that can easily be reduced to the divalent ions, the excitation energy can be transferred from an LMCT state to the 4f levels of the lanthanide ions when the LMCT state lies at high-enough energy. Detailed investigation on the numerous energy transfers processes is very important in order to well tune the luminescent properties of lanthanide MOFs.

In MOFs, efficient lanthanide-centered luminescence is usually fulfilled by the use of antenna linkers, in which, the following conditions have to be considered: (1) The intersystem crossing yield of the antenna linkers should be high, although the lanthanide ion can enhance the intersystem crossing yield via an external heavy atom effect. (2) The linkers should have matching triplet state energy levels for the resonance level of the lanthanide ion. If the energy difference between donor and acceptor is too small, a thermally activated energy back transfer can occur, whereas large energy differences may lead to slower energy transfer rates. The energy of the triplet state must be elaborately tuned to maximize the transfer and minimize the back transfer. (3) The antenna linkers should have a high absorption coefficient, in order to obtain high luminescence intensities. (4) The antenna linkers should be in close proximity to the lanthanide ion, because the energy transfer process is strongly distance dependent.

Thus, it is of great significance to select or design suitable ligand with the appropriate energy level. To illustrate the sensitizations effect of ligands, we show here two representative examples of such luminescent lanthanide MOFs. The first example is the isostructural lanthanide (Sm^{3+}, Eu^{3+}, Gd^{3+}, Tb^{3+}, Dy^{3+}) MOFs containing 4,4′-disulfo-2,2′-bipyridine-*N*,*N*′-dioxide as linker. The triplet state energy levels of ligand was determined to be 24,038 cm^{-1}; thus the ligand is capable of sensitizing the trivalent ions Sm^{3+}, Eu^{3+}, Tb^{3+}, and Dy^{3+} in which strong emissions observed have been observed in Eu^{3+} and Tb^{3+} MOFs. In the second example, the thiophenyl-derivatized carboxylic acid has been selected as possible sensitizing chromophore and multicarboxylate linker to construct the lanthanide MOFs [99]. Because of the big radius of the S atom, its lone pair of electrons can be more easily delocalized within the heterocycle, and the ligand exhibits good charge-transfer ability. The results show that Eu^{3+} and Tb^{3+} MOFs exhibit typical intense luminescence in the visible region, and the Yb^{3+} MOFs emit moderate characteristic near-infrared luminescence in solid state even with coordinated solvents.

In hetero-lanthanide MOFs, a similar energy transfer from ligands to lanthanide occurs through which multiple lanthanide elements may also be excited, but instead of stimulating emission from each, energy can be transferred from one lanthanide to another, resulting in a preferential enhancement of a single lanthanide-centered luminescence. For instance, the MOF $[(Eu, Tb)(C_6H_8O_4)_3(H_2O)_2]\cdot(C_{10}H_8N_2)$

shows the three major peaks at 545, 595, and 615 nm resulting from both Tb^{3+} and Eu^{3+} emission [100]. The peaks corresponding to Eu emission (595 and 615 nm) are more intense than in the isostructural Eu-only analog, and Tb emission at 545 nm is nearly completely quenched, suggesting that the Tb^{3+} center is also acting to sensitize the Eu^{3+} then effectively enhancing Eu^{3+} emission. The similar Tb^{3+} to Eu^{3+} energy transfer has also been observed in the MOFs $(Eu_{0.2}Tb_{0.8})_2$(2,5-pdc)$_2$(1,4-pda)$(H_2O)_2$ and $(Eu_{0.1}Tb_{0.9})_2$(2,5-pdc)$_2$(1,4-pda)$(H_2O)_2$ (H_2pdc = 2,5-pyridinedicarboxylic acid, H_2pda = 1,4-phenylenediacetic acid) [101]. In these MOFs, the 14 K emission spectra excited at the Tb^{3+} $^5D_3$ level (377 nm) show the typical Eu^{3+} and Tb^{3+} lines, clearly supporting the abovementioned Tb^{3+} to Eu^{3+} energy transfer. Excitations at the band from 255 to 310 nm produce similar spectra to those obtained under direct Tb^{3+} excitation, suggesting that ligand to Tb^{3+}, ligand to Eu^{3+}, and Tb^{3+} to Eu^{3+} energy transfer occurs. On the contrary, the spectra excited at the Eu^{3+} $^5D_2$ level (464 nm) only show the typical Eu^{3+} lines.

Up-conversion is the generation of higher-energy photons from lower-energy radiation on the basis of sequential adsorption and energy-transfer steps. Up-conversion luminescent materials have been extensively investigated for their potential applications in laser, display, bioassay, and bioimaging. In particular, NIR-to-visible up-conversion luminescent materials which emit visible light upon NIR excitation have attracted much attention in biomedicine due to the absence of photodamage to live organisms, low autofluorescence background, high signal-to-noise ratio, and high light-penetration depth in biological tissues [102, 103]. Up to now, many kinds of lanthanide ion-doped inorganic materials have been widely investigated for up-conversion luminescence. However, only a few examples of up-conversion luminescent lanthanide MOFs have been reported. Yang et al. prepared a MOF Nd_2(1,4-NDC)$_3$(DMF)$_4$·H_2O (1,4-NDC = 1,4-naphthalenedicarboxylate) [104], which exhibits a weak UV up-conversion emission at about 391.6 nm and a much stronger blue emission at about 449.5 nm upon pulse laser excitation at 580 nm. Weng et al. reported a Y:Er-Yb codoped MOF [(Y:Er-Yb)$_3$(1,4-BDC)$_{3.5}$(OH)$_2$(H_2O)$_2$]·H_2O [105], which exhibits four emission bands of Er^{3+} under 980-nm laser excitation. The red emission attributed to the $^4F_{9/2} \rightarrow {}^4I_{15/2}$ transition is centered at 654 nm, and the green bands ascribed to the transitions of $^4S_{3/2}/^2H_{11/2} \rightarrow {}^4I_{15/2}$ are at 545 and 524 nm, respectively. A rarely observed indigo emission at 455 nm can also be observed, corresponding to the transition $^4F_{5/2} \rightarrow {}^4I_{15/2}$. The 545- and 524-nm emissions are described as a two-photon up-conversion excitation mechanism, and the up-conversion process is also discussed. First, the Yb^{3+} ions are easily excited to their $^2F_{5/2}$ level by absorbing a 980-nm photon, and then the energy is transferred to the Er^{3+} ions which can be excited to the $^4I_{11/2}$ level. A second photon transfer from the excited Yb^{3+} ions pumps the Er^{3+} ions from $^4I_{11/2}$ to the $^4F_{7/2}$ level, which can decay nonradiatively to the $^4S_{3/2}$ and $^2H_{11/2}$ levels. Thus, the green emissions observed result from the $^4S_{3/2} \rightarrow {}^4I_{15/2}$ and the $^2H_{11/2} \rightarrow {}^4I_{15/2}$ transitions. Furthermore, the 455-nm emission is explained through the three-photon up-conversion excitation mechanisms.

The luminescence of Ln^{3+} ion from the f–f transitions can be classified as two types of transitions: the parity-allowed magnetic dipole transitions and the parity-forbidden electric dipole transitions. When the Ln^{3+} ion is inserted into a chemical environment, noncentrosymmetric interactions allow the mixing of electronic states of opposite parity into the 4f wave functions, and electric dipole transitions become partly allowed. The intensity of some of these transitions is particularly sensitive to the nature of the metal ion environment, and these transitions are called hypersensitive transition; a typical example is the $^5D_0 \rightarrow {}^7F_2$ transition of Eu^{3+} [106]. Thus, the luminescence of lanthanide ions can provide valuable information about the local environment and make them very suitable for acting as a structural probe deciphering the symmetry of the chemical environment and the coordination sphere.

Furthermore, the lanthanide-centered emission in MOFs is sensitive to the lowest triplet level of the ligands, suggesting the possibility to modulate the luminescence intensity for a given lanthanide ion by varying the interaction between the ligands and the analyte. These interactions may facilitate or disrupt the energy transfer process, including a modification of the energy transfer ability of the bound ligands or energy transfer from the analyte onto the lanthanide ion, even quenching instead of sensitization. Therefore, the luminescence of lanthanide ions is also potentially useful for a variety of analytical probes. Nowadays, chemical sensors based on lanthanide MOFs for cations, anions, small molecules, gases, and explosives have been designed along these lines.

As remarked above, the luminescence of Eu^{3+} ions is most appropriate as structural probe for the determination of the number of metal ion sites in a compound, their symmetry, their hydration numbers, and their coordination sphere. Especially, the ratio of the intensities of the $^5D_0 \rightarrow {}^7F_1$ and $^5D_0 \rightarrow {}^7F_1$ transitions is very sensitive to the symmetry of the Eu^{3+} centers, because the $^5D_0 \rightarrow {}^7F_1$ emission is due to the magnetic dipole and independent of the environment, while the $^5D_0 \rightarrow {}^7F_2$ emission is due to the electric dipole and is sensitive to the crystal field symmetry. Four examples illustrating the luminescence of Eu^{3+} ions can be utilized as a structural probe in MOFs. In the first example, a Eu-doped MOF $[La_2(H_2O)_4][(C_5H_3N(COO)_2)_2(C_6H_4(COO)_2)]$ [107], the intensity ratio of the $^5D_0 \rightarrow {}^7F_2$ and $^5D_0 \rightarrow {}^7F_1$ transitions is 5.2, and the intensity ratio of dehydrated samples is determined to be 4.2. The fully hydrated MOF contains two water molecules positioned adjacent to each other in one face of the trigonal prism, as part of the nine coordination of the La^{3+} ion, and can exert some strain. Upon dehydration, the bound water molecules are removed, which would give rise to a seven-coordinated La^{3+} ion that is probably much less strained than the nine-coordinated La^{3+} site. Thus, the more symmetric environment of lanthanide ions in the dehydrated MOF results in the decrease of the intensity ratio value.

The second example MOF $Eu_3(2,6\text{-pydc})_3(2,6\text{-Hpydc})(SO_4)(H_2O)_3{\cdot}3(H_2O)_3$ (2,6-pydc = pyridine-2,6-dicarboxylate) [108] crystallizes in the triclinic space group and possesses a 2D metal-organic framework based on hexanuclear Eu_6 SBUs. The asymmetric unit comprises three independent Eu^{3+} ions, three $pydc^{2-}$ dianions, one $Hpydc^{-}$ anion, one sulfate, three coordinated, and three guest water molecules. The intensity ration of the $^5D_0 \rightarrow {}^7F_2$ and $^5D_0 \rightarrow {}^7F_1$ transitions is about 5.2; in addition, the symmetry-forbidden emission $^5D_0 \rightarrow {}^7F_0$ is observed at 579 nm. These phenomena indicate that Eu^{3+} ions occupy sites with low symmetry and have no inversion center.

In the third example, Chandler et al. synthesized a MOF $[Na_6(H_2O)_6][Eu(L)_4](H_2O)_nCl$ ($L = 4,4'$-disulfo-2,2′-bipyridine-*N*,*N*′-dioxide) [109], in which, the metal center exists in a mildly distorted square antiprismatic geometry resulting in a D_4 point symmetry based on the crystallographic data. Generally, when Eu^{3+} ions exist in a D_4 site symmetry, two components of the $^5D_0 \rightarrow {}^7F_1$ and one component of the $^5D_0 \rightarrow {}^7F_2$ transitions are expected to be observed, and the $^5D_0 \rightarrow {}^7F_0$ transition remains strictly forbidden. While the luminescence spectrum displays a weak band ascribed to $^5D_0 \rightarrow {}^7F_0$ transition as well as two peaks of the $^5D_0 \rightarrow {}^7F_1$ and two well-resolved sharp peaks of the $^5D_0 \rightarrow {}^7F_2$ transition, indicating a small distortion from an ideal D_4 symmetry. When the MOF is fully dehydrated via thermal treatment, the emission spectrum remains nearly identical to the original, except the weak band of $^5D_0 \rightarrow {}^7F_0$ transition disappears, suggesting that the water can perturb the ligand system to a minor extent inducing the noncentrosymmetric ligand field.

Finally, the emission spectra and $^5D_0$ lifetime of Eu^{3+} are utilized to estimate the strength of the Ln–ligand bond and the hydration numbers. Shi et al. describe a Eu MOF linked with etidronic acid, named as $Na_4[Eu_2(H_2O)_2(hedp)_2]{\cdot}H_2O$ [110], whose structure comprises eight-membered ring channels filled with Na^+ and both free and Eu-coordinated water molecules. When excited at 250 and 395 nm, the luminescence spectra display the typical Eu^{3+}-centered emission ascribed to the $^5D_0 \rightarrow {}^7F_{0\text{-}4}$ transitions, whose energy and number of Stark components is independent of the excitation wavelength, and only one line for the $^5D_0 \rightarrow {}^7F_0$ transition is observed. This suggests that the Eu^{3+} ions occupy the same average local environment in as-synthesized and dehydrated MOF. Dehydration induces the decrease and red shift for the $^5D_0 \rightarrow {}^7F_0$ transition, indicating that the Eu–O bonds are slightly less covalent in the dehydrated MOF, in accord with the X-ray diffraction evidence. Additionally, the presence of water molecules in the first coordination sphere is estimated from the emission spectra and $^5D_0$ lifetime values. As mentioned above, lanthanide luminescence is very sensitive to the quenching by high-energy vibrations such as O–H. This quenching can be turned into an advantage for assessing the number of water molecules q interacting in the inner coordination sphere from lifetimes measured in water and deuterated water. Assuming that the O–H quenching is the main nonradiative process operating and that all the other deactivation paths are the same in water and in deuterated water, the hydration number can be determined by measuring the lifetime in the deuterated solvent [106]. Thus, the number of water molecules coordinated to Eu^{3+} is determined using an empirical formula, and the results for the as-synthesized and rehydrated samples are very similar, indicating one water molecule in the coordination sphere of Eu^{3+}, in accord with the crystal structure.

3.3 Charge-Transfer Luminescence

Charge-transfer luminescence is an allowed transition from the charge-transfer excited state to the ground state. Ligand-to-metal charge transfer (LMCT) and metal-to-ligand charge transfer (MLCT) states are the excited states commonly found in MOFs. LMCT states involve the electronic transitions from a ligand-localized orbital to a metal-centered orbital, while MLCT states correspond to the electronic transitions from a metal-centered orbital to a ligand-localized orbital. The charge-transfer luminescence is commonly observed in d^{10} metal-based MOFs. For example, in the MOFs $Cu_3(C_7H_2NO_5)_2{\cdot}3H_2O$ ($C_7H_2NO_5$ = 4-hydroxy-pyridine-2,6-dicarboxylate) and $CuAg_2(C_7H_3NO_5)_2$ [111], the former exhibits blue fluorescence with the maximum emission at 398 and 478 nm upon excitation at 333 nm, while the latter and the free ligand 2,6-dicarboxy-4-hydroxypyridine both display green fluorescence with the maximum emission at 515 and 526 nm upon excitation at 358 and 365 nm, respectively. Compared with the ligand, the MOF $Cu_3(C_7H_2NO_5)_2{\cdot}3H_2O$ displays two large blue shift of 59 and 48 nm in luminescent spectra, respectively, while $CuAg_2(C_7H_3NO_5)_2$ displays a little blue shift of 11 nm, suggesting that the emission of the former may be originated from MLCT. The origin of the emission bands are further confirmed by the calculated energy band structures. This indicates that the luminescence of MOF $Cu_3(C_7H_2NO_5)_2{\cdot}3H_2O$ may be ascribed to MLCT where the electrons are transferred from the Cu-3d to O-2p and N-2p states, while the emission band of $CuAg_2(C_7H_3NO_5)_2$ may be originated from π–π^* transition of the ligand.

Luminescence originated from LMCT excited states have been reported in some MOFs. For example, the MOF $[Zn(2,3\text{-pydc})(bpp)]{\cdot}2.5H_2O$ and $[Cd(2,3\text{-pydc})(bpp)(H_2O)]{\cdot}3H_2O$ (2,3-pydcH2 = pyridine-2,3-dicarboxylic acid) display intense fluorescent emissions at 436 and 438 nm upon excitation at 372 and 370 nm [112], respectively, while the ligand 2,3-pydcH2 displays very weak photoluminescent property upon excitation at 370 nm. Thus the origin of the emission for these MOFs can be attributed to LMCT luminescence. In a homochiral Cd MOF $[Cd(dtba)(bpp)]_{3n}$ (H_2dtba = 2,2′-dithiobisbenzoic acid, bpp = 1,3-bis(4-pyridyl)propane) [113], the ligands H_2dtba and bpp both act as multidentate bridging ligands to improve the chain helicity of the product and have the ability to produce unique interwoven extended structural motifs. Interestingly, the luminescence of the MOF exhibits clear difference when excited at room and low temperatures. The MOF displays a blue emission band at 434 nm with a shoulder peak at 482 nm when excited at 355 nm at room temperature, which is attributed to the metal-perturbed intraligand emissions of dtba or bpp. When frozen to 10 K, the emission of the MOF shifts to longer wavelength at about 507 nm and with only one peak, which is attributable to LMCT.

Sometimes, LMCT or MLCT luminescence may compete with ligands-based luminescence, resulting in two emission bands. For example, excitation of the zinc MOF $Zn_2(ATA)_3(ATA)_{2/2}$ at 320 nm at room temperature produces a blue luminescence with a weak emission peak centered at 485 nm that may be assigned to the

LMCT or MLCT and a significantly strong emission with the maximum peak at 392 nm originated from ligands-based emission [114].

3.4 Guest-Induced Luminescence

Serving as host matrices, MOFs can offer a unique platform to incorporate and confine the luminescent species due to their highly regular channel structures and controllable pore sizes. In fact, through the introduction of a variety of metal ions, quantum dots, nanoparticles, and fluorescent dyes into their pore spaces, many types of guest-induced luminescence behaviors have been observed in the MOF composites. An et al. prepared a series of lanthanide ion-doped MOFs Ln^{3+}@bio-MOF-1 ($Ln^{3+} = Tb^{3+}$, Sm^{3+}, Eu^{3+}, or Yb^{3+}) from the as-synthesized bio-MOF-1 via cation exchange process [115]. When excited at 365 nm, the doped MOFs emitted their distinctive colors (Eu^{3+}, red; Tb^{3+}, green; Sm^{3+}, orange pink), which are readily observed with the naked eye. Tb^{3+}@bio-MOF-1, Sm^{3+}@bio-MOF-1, Eu^{3+}@bio-MOF-1, and Yb^{3+}@bio-MOF-1 exhibit the emission at 545, 640, 614, and 970 nm, respectively. It needs to be mentioned that there exists another main emission at 340 nm in all these lanthanide ion-doped MOFs, suggesting that energy migrates through the same electronic levels located in the MOF chromophoric structure for doped lanthanide ions. Notably, the characteristic luminescence of lanthanide ions can even be detected in aqueous environments, despite the strong quenching effect of water molecules, indicating that the bio-MOF-1 scaffold can not only effectively sensitize but also sufficiently protect the lanthanide ions. Furthermore, the quantum yields of lanthanide emission are all reasonably high in aqueous environments, demonstrating that the lanthanide ions are well protected within the pores, and the energy transfers from MOFs to lanthanide ions are efficient. Luo et al. prepared the Eu^{3+}- and Tb^{3+}-doped MOFs which exhibit tunable luminescence properties and sensing functions for metal ions [116].

Except for the lanthanide ions, complex, fluorescent dyes and quantum dots have also been encapsulated into MOFs to generate or tune the luminescence. A mesoporous MOF, $[Tb_{16}(TATB)_{16}(DMA)_{24}]\cdot(DMA)_{91}(H_2O)_{108}$ (H_3TATB = 1,3,5-tribenzoic acid), containing cages of 3.9 and 4.7 nm in diameter [117], emits strong green light originated from Tb^{3+} at 488, 541, 584, and 620 nm. Upon ferrocene is introduced into the framework by a sublimation procedure at 100°C, no strong green emission is observed; instead, a weak and broad emission from the included ferrocene molecules appears. This may be attributed to the efficient energy transfer from the host framework to the ferrocene molecules. Elemental analysis of the material reveals uptake of roughly 65 ferrocene molecules per formula unit or 4,420 per pore. Removal of the ferrocene under vacuum and high temperature leads to recovery of the green emission from the MOF; moreover, the spectrum exactly matched that of the evacuated host crystals, and the ferrocene emission disappeared.

The incorporation of highly luminescent core–shell quantum dots (QDs) within MOF-5 is achieved [118]. Through appropriate surface functionalization, the yellow- and red-emitting QDs are solubilized within MOF-5 growth media. This permits the

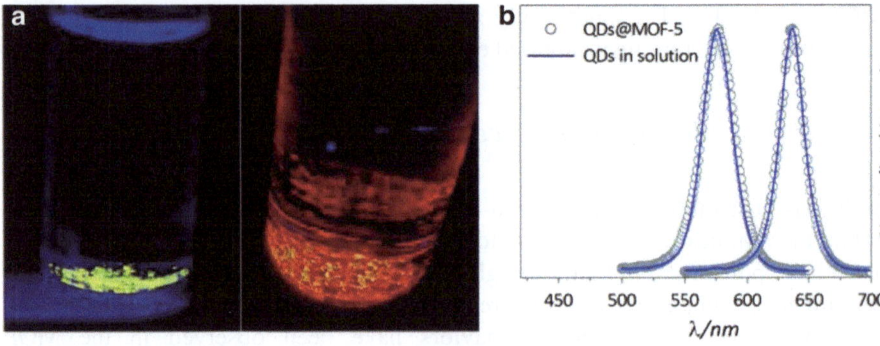

Fig. 3 (**a**) Digital photograph of vials containing yellow (*left*) and red (*right*) QD@MOF-5 samples illuminated with a black light. The samples are kept in dichloromethane. (**b**) Linear emission spectra of yellow and red QDs in solution (*continuous lines*) and of the correspondent QD@MOF-5 samples (*dots*). Reprinted with the permission from [118], Copyright 2012, John Wiley & Sons Ltd.

incorporation of the QDs within the evolving framework during the reaction. The resultant MOF composites exhibit yellow emission centered at 578 nm and the red emission at 637 nm, respectively (Fig. 3). The unchanged emission peak profile of the QDs embedded within the MOF framework suggests that the harsh chemical environment during MOF formation did not degrade the QD core–shell structure.

In addition, the classical fluorescent dye rhodamine 6G (Rh6G) has been encapsulated into a large porous MOF, $Cd_3(bpdc)_3(DMF)]\cdot 5DMF\cdot 18H_2O$ (H_2bpdc = 4,4′-biphenyldicarboxylic acid), reported by Fang et al. Rh6G is a well-known xanthene derivative used as lasers dye, and it exhibits strong absorption in the visible region and a very high fluorescence quantum yield. The MOF Cd $_3(bpdc)_3(DMF)]\cdot 5DMF\cdot 18H_2O$ possesses a 1D hexagonal nanotube-like channels of 24.5×27.9 Å, allowing the entry of the Rh6G dye molecules. The Rh6G-filled MOF exhibits a strongest emission peak at 563 nm originating from the dye Rh6G, which is similar to the previous results obtained on Rh6G-doped materials [119].

Up to date, a variety of luminescent MOFs have been prepared using lanthanide and transition-metal ions. According to the metal ion, three main types of luminescent MOFs can be clearly differentiated: lanthanide-based MOFs, transition-metal-based MOFs, and heterometal-organic frameworks. In addition, a few examples of bismuth-, lead-, and indium-based MOFs have been also reported. The survey of luminescent MOFs and their luminescence wavelength is summarized in Table 1.

4 Applications of Luminescent MOF

As mentioned before, the metal centers, organic moieties, metal-organic charge transfer, and guest molecules within porous MOFs all can potentially generate luminescence. Such uniqueness can allow us to generate luminescent MOF materials with systematically varied emission wavelengths which are crucial for

Table 1 Summary of luminescent MOFs

	MOF	Emission wavelength (nm)Bermudez	References
Lanthanide-based MOFs			
La	$[La(pmtz)(TzC)(H_2O)_3](H_2O)$	530	[120]
	$La_2(PDA)_3(H_2O)_5$	408	[121]
Ce	$[Ce_2(pydc)_2(\mu_4\text{-}SO_4)\cdot 5H_2O]\cdot 2H_2O$	357, 480	[122]
Nd	$Nd_2(1,4\text{-}NDC)_3(DMF)_4\cdot H_2O$	391.6, 449.5	[104]
	$[Nd_4(ox)_4(NO_3)_2(OH)_2(H_2O)_2]\cdot 5H_2O$	~1,060	[123]
	$Nd(tta)_3(\mu\text{-}bpm)\cdot MeOH$	~440, ~910, ~1,070	[124]
	$[Nd(\textit{trans}\text{-}DAM)_2(H_2O)_2]ClO_4\cdot 3H_2O$	912, 1,059, 1,333	[27]
	$[Nd(\textit{trans}\text{-}DAM)_2(H_2O)_2]NO_3\cdot 3H_2O$	912, 1,059, 1,333	[27]
	$[Nd(\textit{trans}\text{-}DAM)(cis\text{-}DAM)(H_2O)_2]Cl\cdot 5H_2O$	912, 1,059, 1,333	[27]
	$[Nd(\textit{trans}\text{-}DAM)_2(H_2O)_2]Cl\cdot 5H_2O$	912, 1,059, 1,333	[27]
	$[Nd(H_2TETA)]NO_3\cdot 2H_2O$	~910, ~1,060, ~1,340	[125]
Sm	$[Sm(pta)(H_2O)_5]\cdot 4H_2O$	558, 592, 639	[126]
	$[Sm_4(pydc)_2(\mu_4\text{-}C_2O_4)_4\cdot 8H_2O]\cdot 6H_2O$	468, 566, 628	[122]
	$Sm_2(ATPA)_3(DMF)_2(H_2O)_2$	450, 561, 597, 644	[127]
Eu	$Eu(1,4\text{-}BDC)_3(H_2O)_4$	592, 617	[128]
	$Eu_2(1,4\text{-}BDC)_3(DMF)_2(H_2O)_2$	596, 616	[129]
	$[Eu(pta)(H_2O)_5]\cdot 4H_2O$	589, 613, 695	[126]
	$[Eu_2(adipic\ acid)_3(H_2O)_2]\cdot 4,4'\text{-}dipyridyl$	595, 615	[130]
	$Eu_3(2,6\text{-}pydc)_3(2,6\text{-}Hpydc)(SO_4)(H_2O)_3\cdot 3(H_2O)_3$	579, 592, 614, 651, 696	[108]
	$Eu(PDC)_{1.5}(DMF)\cdot(DMF)_{0.5}(H_2O)_{0.5}$	590, 616, 698	[131]
	$Eu_2(FMA)_2(OX)(H_2O)_4\cdot 4H_2O$	591, 616, 650, 692	[132]
	$Eu(BTC)(H_2O)\cdot 1.5H_2O$	590, 616, 698	[133]
	$[Eu_2(\mu_{2\text{-}}pzdc)(\mu_4\text{-}pzdc)(\mu_2\text{-}ox)(H_2O)_4]\cdot 8H_2O$	591, 614, 649, 695	[134]
	$[Eu_4(BPT)_4(DMF)_2(H_2O)_8]\cdot(DMF)_5\cdot(H_2O)_3$	580, 650, 592, 617, 700	[135]
	$[Eu_2(fumarate)_2(oxalate)(H_2O)_4]\cdot 4H_2O$	591, 617,	[136]
	$Eu_2(1,4\text{-}BDC)_3(H_2O)_2\cdot(H_2O)_2$	590, 617, 698	[75]

(continued)

Table 1 (continued)

	MOF	Emission wavelength (nm)Bermudez	References
	ITQMOF-1-Eu	619	[137]
	ITQMOF-3-Eu	579.0, 580.7, 612.3	[138]
	$Eu_2(2{,}5\text{-pdc})_2(1{,}4\text{-pda})(H_2O)_2$	~590, 615, ~698	[101]
	$Na_4[Eu_2(H_2O)_2(hedp)_2]\cdot H_2O$	~590, 613	[110]
	$Eu_2(1{,}4\text{-BDC})_3(MeOH)_4]\cdot 8MeOH$	~590, ~615	[139]
	$[Eu(H_2TETA)]NO_3\cdot 2H_2O$	579, 592, 613, 651, 698	[125]
	$Eu_2(Hbidc)_2(ox)_2\cdot(H_2O)_3$	536, 594, 617, 651, 695	[140]
	$Eu_2(ATPA)_3(DMF)_2(H_2O)_2$	450, 616	[127]
Gd	$[Gd_2(Hpimda)_2(\mu_4\text{-}C_2O_4)\cdot 4H_2O]\cdot 2H_2O$	319, 476	[122]
	$[Gd_2(pyda)(\mu_4\text{-}C_2O_4)_2\cdot 4H_2O]\cdot 3H_2O$	299, 546	[122]
Tb	$Tb(1{,}4\text{-BDC})_3(H_2O)_4$	491, 546	[128]
	$[Tb(pta)(H_2O)_5]\cdot 4H_2O$	491, 546, 584, 623	[126]
	$K_5[Tb_5(IDC)_4(ox)_4]$	489, 545, 589, 619	[141]
	$[Tb(Mucicate)_{1.5}\cdot(H_2O)_2]\cdot 5H_2O$	~490, ~545	[142]
	Tb(BTC)·methanol	492, 548, 584, 620	[143]
	$TbNTA\cdot H_2O$	489, 547, 583, 625	[144]
	$Na[Tb(OBA)_2]_3\cdot 0.4DMF_3\cdot 1.5H_2O$	545	[145]
	$[Tb_4(BPT)_4(DMF)_2(H_2O)_8]\cdot(DMF)_5\cdot(H_2O)_3$	490, 544, 587, 621, 650	[135]
	$[Tb_2(fumarate)_2(oxalate)(H_2O)_4]\cdot 4H_2O$	494, 545, 587, 621	[136]
	$Tb_2(2{,}5\text{-pdc})_2(1{,}4\text{-pda})(H_2O)_2$	544, ~585, ~620	[101]
	$Tb(4\text{-hydroxybenzenesulfonate})_3(H_2O)_2\cdot 2H_2O$	~490, ~545	[146]
	$Tb_3(1{,}4\text{-BDC})_{4.5}(DMF)_2(H_2O)_3\cdot(DMF)(H_2O)$	370, 430, 490, ~550, ~590	[147]
	$Tb(1{,}4\text{-BDC})(MeOH)_4]\cdot Cl\cdot MeOH\cdot 0.25H_2O$	~495, ~550	[139]
	$[Tb(H_2TETA)]NO_3\cdot 2H_2O$	492, 545, 584, 620	[125]
	$[Tb_2(Hbidc)_2(ox)(H_2O)_2]\cdot 4H_2O$	490, 548, 590, 620	[140]
	$Tb_2(ATPA)_3(DMF)_2(H_2O)_2$	450, 489, 545, 587, 619	[127]
Dy	$Dy(1{,}4\text{-BDC})_3(H_2O)_4$	481, 575	[128]
	$[Dy(pta)(H_2O)_3]\cdot 4H_2O$	476, 569	[126]

	$K_5[Dy_5(IDC)_4(ox)_4]$	479, 573	[141]
	$[Dy_4(BPT)_4(DMF)_2(H_2O)_8]\cdot(DMF)_5\cdot(H_2O)_3$	480, 573, 662	[135]
	$Dy_2(ATPA)_3(DMF)_2(H_2O)_2$	450	[127]
Er	$[Er(pta)(H_2O)_3]\cdot 4H_2O$	1,538	[126]
	$Er_2(1,4\text{-}BDC)_3(DMF)_2(H_2O)_2\cdot H_2O$	1,550	[148]
	$Er_2(BDC\text{-}F_4)_3(DMF)(H_2O)\cdot DMF$	1,550	[148]
Yb	$[Yb(pta)(H_2O)_3]\cdot 4H_2O$	980	[126]
	$Yb(BPT)(H_2O)\cdot(DMF)_{1.5}(H_2O)_{1.25}$	980	[149]
	$[Yb_2(C_{26}H_{20}O_6)_3(H_2O)_2]\cdot(DMF)_6(H_2O)_{8.5}$	980	[150]
	$[Yb_2(C_{26}H_{20}O_6)_3(H_2O)_2]\cdot(DMF)_{12}(H_2O)_{10}$	980	[150]
	$(Yb_2(TDC)_3(DMF)(H_2O)\cdot(CH_3CH_2OH)$	980	[151]
Transition-metal-based MOFs			
Zn	$[Zn_4O(NTB)_2]\cdot 3DEF\cdot EtOH$	433	[152]
	$Zn_4O(1,4\text{-}BDC)_3$	397	[153]
	$Zn_4(OH)_2(1,2,4\text{-}BTC)_2$	428	[154]
	$(Zn_3(TAA)_3(H_2O)_3\cdot(1,4\text{-dioxane})$	438	[155]
	$[Zn_4O(1,4\text{-}BDC)(bpz)_2]\cdot 4DMF\cdot 6H_2O$	470	[156]
	$Zn_2(bpdc)_2(bpee)$	420	[157]
	Zn–BCPA	438	[158]
	$Zn_3(\mu_5\text{-}pta)_2(\mu_2\text{-}H_2O)_2$	467	[92]
	$Zn_3(BTC)_2(DMF)_3(H_2O)\cdot(DMF)(H_2O)$	410	[93]
	$Zn_3(4,4'\text{-stilbene dicarboxylic acid})_3(DMF)_2$	396, 420, 441	[159]
	$[Zn(2,3\text{-}pydc)(bpp)]\cdot 2.5H_2O$	436	[112]
	$Zn_2(ATA)_3(ATA)_{2/2}$	485	[114]
	$Zn(hfipbb)\cdot 0.5H_2O\cdot 0.5DMF$	425	[160]
	$Zn(hfipbb)(4,4'\text{-}bipy)\cdot DMF$	425	[160]
	$Zn_5(\mu 3\text{-}OH)_2(BTA)_2(tp)_3$	348	[161]
	$Zn(BTA)(chdc)_{0.5}$	380	[161]
	$Zn(BTA)(ap)_{0.5}$	360	[161]
	$Zn(BTA)(gt)_{0.5}$	358	[161]

(continued)

Table 1 (continued)

	MOF	Emission wavelength (nm)Bermudez	References
	Zn(DFDA)	405, 422	[162]
	Zn(DFDA)(1,1′-(1,4-butanediyl)bis(imidazole))	405, 427	[162]
	Zn(DFDA)(2,2′-bipyridine)	411, 427	[162]
	Zn_2(3-tzba)(N_3)(OH)(2,2′-bipy)	420	[163]
	[Zn(3-tzba)(2,2′-bipy)(H_2O)]·3H_2O	437, 465	[163]
	[Zn_2(3-tzba)$_2$(phen)$_2$]·H_2O	427	[163]
	Zn(sfdb)(bpy)(H_2O)]·0.5CH_3OH	359	[164]
	Zn_3(CTC)$_2$(bipy)·(DMF)(H_2O)$_2$	447, 552	[165]
	Zn_3(CTC)$_2$(Bpe)·(DMF)(H_2O)$_2$	456, 560	[165]
Cd	Cd(BDC–NH_2)(bpy)·4H_2O·2.5DMF	435	[166]
	Cd_4(BTC)$_3$(DMF)$_2$(H_2O)$_2$·6H_2O	405	[93]
	[Cd(2,3-pydc)(bpp)(H_2O)]·3H_2O	438	[112]
	[Cd(dtba)(bpp)]$_{3n}$	434, 482	[113]
	Cd(hfipbb)(DMF)·0.5DMF	425	[160]
	Cd(5MT)$_2$	406	[167]
	Cd_5(N_3)(5MT)$_9$·0.12H_2O	405	[167]
	Cd_3(OH)$Cl_{1.39}$(N_3)$_{0.61}$(5MT)$_3$	412	[167]
	Cd_3(OH)Cl(N_3)(ATA)$_3$	422	[167]
	Cd_2(OH)Br(ATA)$_2$	421	[167]
	Cd_7Cl_2(5BT)$_{12}$(H_2O)$_2$	394	[167]
	Cd_7Br_2(5BT)$_{12}$(H_2O)$_2$	395	[167]
	Cd_2(BPVIC)$_2$(SCN)$_4$	562	[168]
	Cd(DFDA)(C_2H_5OH)	418	[162]
	Cd_2(DFDA)$_2$(bis(imidazol-1-ylmethyl)benzene)$_2$	407, 428	[162]
	[Cd(sfdb)(phen)$_2$]·2H_2O	380	[164]
	[Cd(sfdb)(bpy)$_2$]·H_2O	396	[164]
	Cd(sfdb)(quin)	385	[164]
	Cd_3(sfdb)$_2$(Hsfdb)$_2$(phen)$_2$	394	[164]

	Cd(sfdb)(bpy)	365	[164]
	$Cd_2(ABTC)(DMF)_3{\cdot}(DMF)_2$	402	[169]
	$Cd_3(CTC)_2(TED)(H2O)_2{\cdot}(H_3O)_2Cl_2$	522	[165]
	$Cd_3(CTC)_2(bipy)(DMF)_2{\cdot}(DMF)(H_2O)_2$	449, 550	[165]
	$Cd_3(CTC)_2(Bpe)(DMF)_2{\cdot}(DMF)(H_2O)_2$	453, 556	[165]
Ag	$[Ag(4\text{-cyanobenzoate})L]{\cdot}H_2O$	427, 513, 566, 617	[170]
	$[Ag_8(L)_4](NO_3)_8{\cdot}4H_2O$, L = bis(3,5-bis((1H- imidazol-1-yl)-methyl)-2,4,6-trimethylphenyl) methane	495	[171]
	$[Ag_2(L)_2(ClO_4)_2(CH_3CN)]$, L = *N*,*N*′-bis(pyridin-2-ylmethylene)benzene-1,4-diamine	451, 521	[172]
	$[Ag_2(L)_2(ClO_4)_2]$, L = 3,3′-dimethyl-N,N′-bis(pyridin-2-ylmethylene)biphenyl-4,4′-diamine	409, 432, 505, 526	[172]
	$[AgL]ClO_4$, L = *N*′,*N*′-bis[1-(pyridin-4- yl)methylidene]benzil dihydrazone	340	[173]
	$[AgL_2]BF_4$, L = *N*′,*N*′-bis[1-(pyridin-4- yl)methylidene]benzil dihydrazone	342	[173]
	$[AgL(NO_3)](CH_3CH_2OH)$, L = *N*′,*N*′-bis[1-(pyridin-4- yl)methylidene]benzil dihydrazone	339	[173]
	$[AgL(NO_3)]$, L = *N*′,*N*′-bis[1-(pyridin-3- yl)methylidene]benzil dihydrazone)	344	[173]
	$[Ag_2(PhPPy_2)_2Cl](ClO_4)$	~510	[174]
	$Ag(MES)(L)_{0.5}$, L = 1,1′-(1,4- butanediyl)bis(2-methylbenzimidazole)	479	[175]
	$[Ag(MES)(L)_{0.5}]{\cdot}H_2O$, L = 1,1′-(1,4-butanediyl)bis(2-ethylbenzimidazole)	363	[175]
Cu	$Cu_6(5,6\text{-diphenyl-1,2,4-triazine-3-thiol})_6{\cdot}(H_2O)(DMSO)$	660	[176]
	$Cu_3(L)_2{\cdot}3H_2O$, L = chelidamic acid	398, 478	[111]
	$Cu_5(SCN)_5(3\text{-Abpt})_2$	559	[177]
	Cu(SCN)(3-Abpt)	570	[177]
	$Cu_2(SCN)_2(4\text{-PyHBIm})$	569	[177]
	$Cu_2(SCN)_2(3\text{-PyHBIm})$	530	[177]
	Cu(SCN)(4-Ptz)	435	[177]
	$Cu_2(SCN)_2(2\text{-PyBIm})(2\text{-PyHBIm})$	440	[177]
	$[Cu_{12}Br_2(CN)_{6/2}(SCH_3)_6][Cu(SCH_3)_2]$	575	[178]
	$Cu_2I_2(DABCO)_2$	551	[43]
	$Cu_2I_2(piperazine)_2$	560	[43]
	$Cu_2(CN)_2(bpzm)$	473	[179]
	$Cu_2(CN)_2(bpze)$	529	[179]

(continued)

Table 1 (continued)

	MOF	Emission wavelength (nm)Bermudez	References
	$Cu_2(CN)_2$(bpzp)	515	[179]
	$[Cu(Pz)]_3$	542	[180]
	$[Cu_2(Bpz)]_n$	598	[180]
	$Cu_2(ABTC)(H_2O)_2\cdot(DMF)_2(H_2O)$	392	[169]
Co	Co(L)(2,2′-bipy), L = dibenzothiophene- 5,5′-dioxide-3,7-dicarboxylate	409	[181]
Mn	Mn(Hbidc)	726	[182]
	$[MnL_2(NCS)_2]\cdot(CH_3CH_2OH)$, L = *N′*,*N′*-bis[1- (pyridin-3-yl)methylidene]benzil dihydrazone)	356	[173]
Fe	$[FeL_2Cl_2]\cdot CH_3CN$, L = *N′*,*N′*-bis[1-(pyridin-3-yl) methylidene]benzil dihydrazone)	353	[173]
Heterometal-organic frameworks			
Pr–Cu	$Pr(pydc)_3Cu_3(bipy)_3\cdot m(H_2O)$	488, 532	[183]
Nd–Cd	$[NdCd(imdc)(SO_4)(H_2O)_3]\cdot 0.5H_2O$	430	[184]
Nd–Cu	$Nd(pydc)_3Cu_3(bipy)_3\cdot 5(H_2O)$	895, 1,062, 1,345	[183]
Sm–Ag	$[SmAg(PDA)_2(H_2O)_3]\cdot 3H_2O$	563, 601, 644	[185]
Sm–Cu	$Sm(pydc)_3Cu_3(bipy)_3\cdot 4(H_2O)$	563, 603, 645	[183]
	SmCu(nds)(isonicotinic acid)$_2\cdot H_2O$	561, 599, 645	[186]
Sm–Ba	$Ba_2(H_2O)_4[SmL_3(H_2O)_2](H_2O)_nCl$, L = 4,4′-disulfo-2,2′- bipyridine-*N*,*N′*-dioxide	~605, ~640	[187]
Eu–Mn	$[Eu(PDA)_3Mn_{1.5}(H_2O)_3]\cdot 3.25H_2O$	~590, 618	[188]
Eu–Fe	$[Eu(PDA)_3Fe_{1.5}(H_2O)_3]\cdot 1.5H_2O$	613	[189]
Eu–Ag	$[EuAg(PDA)_2(H_2O)_3]\cdot 3H_2O$	581, 593, 617, 651, 695	[185]
	$EuAg(pydc)_2(Hnic)_{0.5}$	~580, ~615	[190]
	$EuAg(inic)_2(nicO)\cdot 0.5H_2O$	~580, ~620	[190]
Eu–Ba	$Ba_2(H_2O)_4[EuL_3(H_2O)_2](H_2O)_nCl$, L = 4,4′-disulfo-2,2′- bipyridine-*N*,*N′*-dioxide	~615	[187]
Eu–Cd	$[EuCd(imdc)(SO_4)(H_2O)_3]\cdot 0.5H_2O$	~590, ~618	[184]
Eu–Cu	$Eu(pydc)_3Cu_3(bipy)_3\cdot 4(H_2O)$	579, 593, 615, 651	[183]
	$[EuCu(nic)_2(ox)]\cdot 2H_2O$	579, 591, 594, 616, 618	[191]
	EuCu(nds)(isonicotinic acid)$_2\cdot H_2O$	579, 592, 619, 654, 704	[186]
Gd–Cd	$[GdCd(imdc)(SO_4)(H_2O)_3]\cdot 0.5H_2O$	355, 400–465, 544	[184]

Gd–Cu	$Gd(pydc)_3Cu_3(bipy)_3 \cdot 4(H_2O)$	488, 532	[183]
Gd–Mn	$[(Gd_2L)Mn(H_2O)_6] \cdot 0.5(H_2O)$, L = 3,3′-(4-Amino-4*H*-1,2,4-triazole-3,5-diyl)dibenzoic acid	428	[192]
Gd–Ba	$Ba_2(H_2O)_4[GdL_3(H_2O)_2](H_2O)_nCl$, L = 4,4′-disulfo-2,2′- bipyridine-*N*,*N*′-dioxide	416, 435, 448	[187]
Tb–Mn	$[Tb(PDA)_3Mn_{1.5}(H_2O)_3] \cdot 3.25H_2O$	~490, ~550	[188]
Tb–Ag	$[TbAg(PDA)_2(H_2O)_3] \cdot 3H_2O$	490, 545, 584, 622	[185]
Tb–Cd	$[TbCd(imdc)(SO_4)(H_2O)_3] \cdot 0.5H_2O$	490, 544, 584, 620	[184]
Tb–Cu	$Tb(pydc)_3Cu_3(bipy)_3 \cdot 4(H_2O)$	492, 544, 583, 622	[183]
	$TbCu(nds)(isonicotinic\ acid)_2 \cdot H_2O$	489, 544, 590, 619	[186]
Tb–Ba	$Ba_2(H_2O)_4[TbL_3(H_2O)_2](H_2O)_nCl$, L = 4,4′-disulfo-2,2′- bipyridine-*N*,*N*′-dioxide	~490, ~545	[187]
Dy–Ag	$[DyAg(PDA)_2(H_2O)_3] \cdot 3H_2O$	483, 573	[185]
Dy–Cd	$[DyCd(imdc)(SO_4)(H_2O)_3] \cdot 0.5H_2O$	478, 573, 657	[184]
Dy–Ba	$Ba_2(H_2O)_4[DyL_3(H_2O)_2](H_2O)_nCl$, L = 4,4′-disulfo-2,2′- bipyridine-*N*,*N*′-dioxide	~640	[187]
Ho–Ag	$HoAg_5(1,2\text{-}BDC)_4$	995, 1,400–1,600	[193]
Er–Ag	$ErAg_5(1,2\text{-}BDC)_4$	1,450–1,650	[193]
	$ErAg_3(L)_3(H_2O)$, L = pyridine-2,6-dicarboxylic acid	1,548	[194]
	$ErAg_3(L)_2(H_2O)$, L = 4-hydroxylpyridine- 2,6-dicarboxylic acid	1,540	[194]
Er–Cu	$Er(pydc)_3Cu_3(bipy)_3 \cdot 4(H_2O)$	488, 532	[183]
Yb–Cd	$[YbCd(imdc)(SO_4)(H_2O)_3] \cdot 0.5H_2O$	433	[184]
Yb–Ag	$YbAg_5(1,2\text{-}BDC)_4$	950–1,050	[193]
Yb–Cu	$Yb(pydc)_3Cu_3(bipy)_3 \cdot 4(H_2O)$	980	[183]
Eu–Ce	$[Eu(dipicH)(H_2O)_6][Ce(dipic)_3]_3 \cdot 7H_2O$	~590, ~615	[195]
Eu–Tb	$Eu_{1\text{-}x}Tb_x(BTC)(H_2O)$	540, 589, 615	[196]
	$Tb(1,3,5\text{-}BTC)(H_2O) \cdot 3H_2O{:}Eu^{3+}$	487, 543, 580, 620	[197]
	$[(Eu, Tb)(C_6H_8O_4)_3(H_2O)_2] \cdot (C_{10}H_8N_2)$	545, 595, 615	[100]
	$(Eu_{0.2}Tb_{0.8})_2(2,5\text{-}pdc)_2(1,4\text{-}pda)(H_2O)_2$	~545, ~595, ~620	[101]
	$Eu_{0.0069}Tb_{0.9931}$-DMBDC	489, 545, 592, 613	[198]
Dy–Ce	$[Dy(dipicH)(H_2O)_6][Ce(dipic)_3]_3 \cdot 7H_2O$	~480, ~575	[195]
Er–Yb	$Er_xYb_{1\text{-}x}$-PVDC-1	980, 1,530	[199]
Zn–Ir	$[Zn_4(\mu_4\text{-}O)(Ir(2\text{-}pyridyl\text{-}benzoic\ acid)_3)_2] \cdot 6DMF \cdot H_2O$	538	[200]

(continued)

Table 1 (continued)

	MOF	Emission wavelength (nm)Bermudez	References
Cu–Ag	$CuAg_2(L)_2$, L = chelidamic acid	515	[111]
Others			
In	$In_2(OH)_2(TBAPy)$	471	[201]
	$[In_3(BTB)_2(ox)_3]_n$	~460	[202]
Bi	$Bi_3(\mu_3\text{-}O)_2(pydc)_2(Hpydc)(H_2O)_2$	430, 460, 480, 556	[203]
Pb	$Pb(pydc)(H_2O)$	441, 470, 520, 563	[203]
	$Pb_4(1,3\text{-}BDC)_3(\mu_4\text{-}O)(H_2O)$	424	[204]
	Pb(bpdc)	480	[205]

the lighting, display, and optical devices. The robust pores within luminescent MOFs are particularly useful to develop luminescent sensing materials. In fact, to make use of the pores within luminescent MOFs for their differential recognition of sensing substrates, a number of porous luminescent sensing MOFs have been fulfilled. Luminescent MOFs might also be utilized as light-emitting materials for white-light and near-infrared light-emitting devices. Nanoscale luminescent MOFs have started to be explored for their applications on tissue and cell imaging, as well as drug delivery monitoring and treatment.

4.1 Near-Infrared (NIR) Luminescent Materials

Recent researches in the field of lanthanide-based near-infrared (NIR) luminescent materials are very active due to their advantages for lasers, optical amplifiers for telecommunication, biological applications, NIR organic light-emitting diode (OLED) technology, solar energy conversion, etc. For example, lanthanide ions Nd^{3+} and Er^{3+} exhibit near-infrared luminescence at wavelengths of 1,330 and 1,550 nm, respectively, which are of interest for applications in optical telecommunication networks. However, a few examples of NIR light-emitting lanthanide frameworks are reported in comparison to visible-emitting MOFs due to their smaller energy gap between the ground state and emissive state of these ions, as compared to Eu^{3+} and Tb^{3+}, thus resulting in more efficient nonradiative decay through high-energy C–H, N–H, and O–H oscillators in the ligands and solvents [96].

In order to alleviate nonradiative decays, a variety of strategies have been adopted to shield the ion excited levels to high nonradiative transition probability by O–H, C–H, and N–H oscillators. Chen et al. demonstrated that the NIR luminescence intensity of MOFs can be significantly enhanced by incorporation of fluorinated organic linkers. They synthesized two erbium–organic frameworks $Er_2(1,4\text{-BDC})_3(DMF)_2(H_2O)_2{\cdot}H_2O$ (1,4-BDC = 1,4-benzenedicarboxylate) and $Er_2(BDC\text{–}F_4)_3(DMF)(H_2O){\cdot}DMF$ ($BDC\text{–}F_4$ = 2,3,5,6-tetrafluoro-1,4-benzenedicarboxylate) [148]. After heat treatment under vacuum at 140°C overnight, the desolvated framework displays NIR emission band around 1.55 μm assigned to the $^4I_{13/2} \rightarrow {}^4I_{15/2}$ transition of Er^{3+} ion at the excitation wavelength of 808 nm. Interestingly, the emission intensity of $Er\text{–}BDC\text{–}F_4$ framework is significantly enhanced by 3 times than that of Er–BDC framework because fluorination can significantly reduce the fluorescence quenching effect of the vibrational C–H bond. Chen et al. speculated that the luminescent intensity of the fluorinated framework can be further enhanced by incorporating ancillary electron-rich organic ligands instead of solvent molecules to saturate the Er^{3+} coordination sites.

Additionally, a series of NIR luminescent Ln–Ag HMOFs, $LnAg_5(1,2\text{-BDC})_4$ (Ln = Yb, Er, Ho, 1,2-BDC = 1,2-benzenedicarbolylate) have been reported by Jin et al. [193]. The HMOF $YbAg_5(1,2\text{-BDC})_4$ shows a broad emission band ranging from 950 to 1,050 nm assigned to the $^2F_{5/2} \rightarrow {}^2F_{7/2}$ transition of the Yb^{3+} ion, and the $ErAg_5(1,2\text{-BDC})_4$ exhibits a broad emission band extending from 1,450 to 1,650 nm attributed to the $^4I_{13/2} \rightarrow {}^4I_{15/2}$ transition of an Er^{3+} ion. The $HoAg_5(1,2\text{-}BDC)_4$ displays an emission at 995 nm assigned to the $^5F_5 \rightarrow {}^5I_7$ transition of Ho^{3+} ion and a band from 1,400 to 1,600 nm corresponding to the splitting of $^5F_5 \rightarrow {}^5I_6$ transition. Especially, the NIR emission for these HMOFs show shift, splitting or broadening emission spectra compared with the isolated lanthanide ions, which might be attributed to the interaction and influence of 4d and 4f orbitals. Jin et al. speculated that the Ag–Ln distance becomes much shorter (ranging from 3.71 to 3.75 Å) upon formation of the Ag–Ln HMOFs, so the 4d orbital of the Ag^+ ion and 4f orbital of the Ln^{3+} ions may interact and influence each other, resulting in the tuned inner levels of the Ag–Ln system and shifting and/or split of the lower-energy states of Ln^{3+} ions. Other NIR luminescent HMOFs include an infinite triple-stranded helical framework $ErAg_3(L)_3(H_2O)$ (L = pyridine-2,6-dicarboxylic acid) and a sandwich-type 3D framework $ErAg_3(HO\text{–}L)_2(H_2O)$ (HO–L = 4-hydroxylpyridine-2,6-dicarboxylic acid) [194]. The HMOF $ErAg_3(L)_3(H_2O)$ shows a broad emission band at 1,548 nm ascribed to the $^4I_{13/2} \rightarrow {}^4I_{15/2}$ transition of Er^{3+} ion. Similarly, $ErAg_3(HO\text{–}L)_2(H_2O)$ exhibits an emission band with the strongest peak at 1,540 nm.

Several examples of Ln–Cu HMOFs with mixed bridging ligands Ln $(pydc)_3Cu_3(bipy)_3{\cdot}m(H_2O)$ (Ln = Pr, Nd, $m = 5$; Ln = Sm, Eu, Gd, Tb, Er, Yb, $m = 4$; pydc = 2,6-pyridinedicarboxylate anion; bipy = 4,4′-bipyridine) were reported by Bo et al. [183]. During the self-assembly process, the heterometallic ions are interconnected by mixed ligands to produce spindle-shaped heterometallic rings $[Ln_6(pydc)_6Cu_{12}(bipy)_6]$ as the SBUs which are pillared by bridging bipy molecules to yield the porous 3D pillared-layer Ln–Cu HMOFs. The Nd–Cu HMOFs display the strongest emission band at 1,062 nm ascribed to $^4F_{3/2} \rightarrow {}^4I_{11/2}$ transition of Nd^{3+} ion, a broad emission band centered at 895 nm attributed to $^4F_{3/2} \rightarrow {}^4I_{9/2}$ transition, and a sharp weak emission at 1,345 nm corresponding to $^4F_{3/2} \rightarrow {}^4I_{13/2}$ emission. The Yb–Cu HMOF also displays a strong emission at 980 nm attributed to the $^2F_{5/2} \rightarrow {}^2F_{7/2}$ transition of the Yb^{3+} ion in near-infrared region upon the excitation at 275 nm. All Sm–Cu, Eu–Cu, and Tb–Cu HMOFs exhibit the characteristic emission of the corresponding lanthanide ions; while the Pr–Cu, Gd–Cu, and Er–Cu HMOFs display two intense emission peaks at 488 and 532 nm, similar to those found in Cu(I) complexes, which are assigned to the emission from MLCT transition.

Although some excellent ligands with high intersystem crossing yields such as β-diketonates and triphenylene derivative have been designed for sensitizing lanthanide ions; a drawback of these sensitizers is that they require near-UV excitation and that they have relatively high-lying triplet energy levels. White et al. utilized a new ligand to efficiently sensitize the NIR emission in the Yb^{3+}-MOFs. Based on the ligand 4,4′-[(2,5-dimethoxy-1,4-phenylene)di-2,1-ethenediyl]bisbenzoic acid (H_2-PVDC), two MOFs, named $[Yb_2(PVDC)_3(H_2O)_2]{\cdot}(DMF)_6(H_2O)_{8.5}$ (Yb-PVDC-1) and $[Yb_2(PVDC)_3(H_2O)_2]{\cdot}(DMF)_{12}(H_2O)_{10}$ (Yb-PVDC-2) were synthesized [150]. These MOFs display the NIR emission at 980 nm of Yb^{3+} ion, but the excitation band of Yb-PVDC-2 is red shifted to 500 nm from 470 nm in Yb-PVDC-1, which was proposed that the close π–π interactions between the PVDC linkers decrease the energy of the $\pi \rightarrow \pi^*$ transition, resulting in a lowered excitation energy. Additionally, the quantum yield of Yb-PVDC-2 is five times

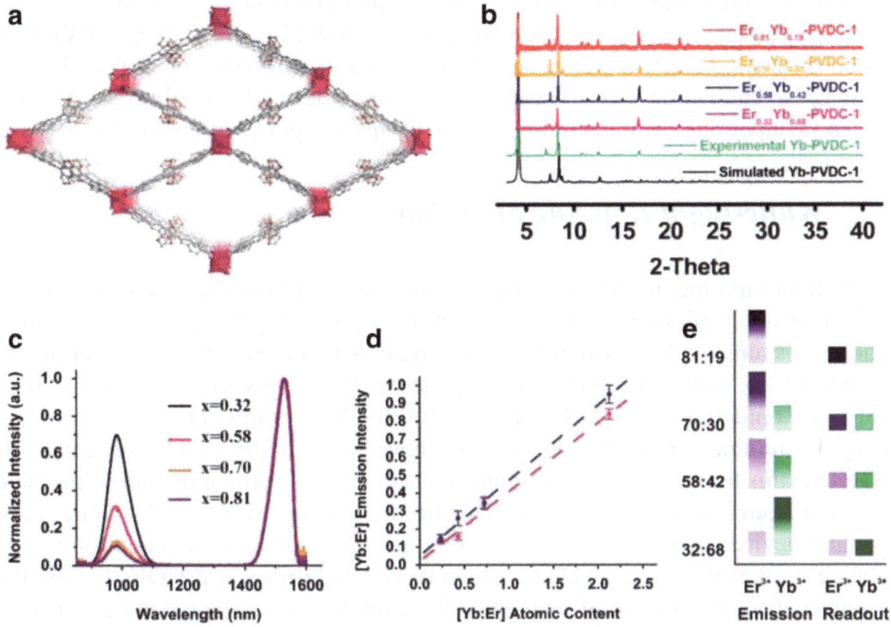

Fig. 4 (**a**) Crystal structure of Yb-PVDC-1 viewed along the crystallographic *c* axis. (**b**) PXRD patterns for Yb-PVDC-1 and analogs Er_xYb_{1-x}-PVDC-1 ($x = 0.32, 0.58, 0.70$, and 0.81). (**c**) Yb^{3+} (980 nm) and Er^{3+} (1,530 nm) emission spectra of Er_xYb_{1-x}-PVDC-1 ($x = 0.32, 0.58, 0.70$, and 0.81) normalized to the Er^{3+} signal upon 490-nm excitation. (**d**) Plot of the ratio of integrated emission intensities versus their atomic ratio. (**e**) Color-coded schematic of the barcode readout. Reprinted with the permission from [199]. Copyright 2009, American Chemical Society

higher than that of Yb-PVDC-1 when excited through the lower energy band (490 nm). The quantum yield is among the highest values reported for Yb^{3+} systems under solvent, which may be attributed to the different coordination environments and levels of protection in both MOFs. This example demonstrates a new route for controlling the luminescent properties of lanthanide MOFs without altering the structure of the sensitizer.

Furthermore, White et al. proposed a new approach for creating luminescent barcoded systems by doping multiple Ln^{3+} ions into the abovementioned MOFs, which simultaneously emit several independent NIR signals arising from different lanthanide compositions [199]. They designed Er_xYb_{1-x}-PVDC-1 ($x = 0.32, 0.58$, 0.70, and 0.81) MOFs with different amounts of Yb^{3+} and Er^{3+} ions that were sensitized by the same ligand PVDC. These MOFs displayed sharp signals from both Er^{3+} (1,530 nm) and Yb^{3+} (980 nm), and the excitation spectrum of either the Er^{3+} or Yb^{3+} emission band contains two similar bands with apparent maxima at 370 and 470 nm. By controlling reactant stoichiometry, the lanthanide composition of the MOF and the resulting individual emission intensities can be tuned and controlled (Fig. 4). This features their potential usage as NIR barcodes. Furthermore, the number and diversity of barcodes may be increased by using a large number of Ln/Ln ratios or by incorporating additional lanthanide ions into the material. The later concept was demonstrated in MOF $Nd_{0.09}Er_{0.55}Yb_{0.36}$-PVDC-1, which displays a more sophisticated barcode signal consisting of NIR emissions from three lanthanide ions. These MOFs retain their NIR luminescence when coated in superglue, demonstrating the possibility for practical application.

4.2 White-Light-Emitting Materials

White-light-emitting materials and devices have attracted significant attention due to their broad applications in displays and lighting. High-quality white-light illumination requires a source with the Commission International ed'Eclairage (CIE) coordinates (0.333; 0.333), with correlated color temperature (CCT) between 2,500 and 6,500 K, and color rendering index (CRI) above 80 [206, 207]. Emission from organic or inorganic luminescent materials can only cover part of the visible spectrum. To overcome this limitation, various architectures of devices combining monochromatic emission from different compounds have been suggested. In principle, white-light-emitting devices may be obtained by tuning the emission color and controlling the relative amount of these monochromatic emissions. In luminescent MOFs, besides the characteristic f–f emission of lanthanide ions, the broad emission band ascribed to organic linkers can be observed. The relative luminescence intensity between the lanthanide ions and linkers strongly depends on the nature of the energy transfer, thus allowing the fine-tuning of the MOFs emission color across the CIE diagram. In addition, the emission color can also be readily modulated both by chemical factors (Ln^{3+} types and concentration, ligand structure, coordination status, guest species) and physical parameters (excitation wavelength and temperature).

Doping lanthanide ions into the isostructural MOF materials has been demonstrated as an effective approach to tune the emission colors. Qian and Chen et al. reported a new two-dimensional lanthanide MOF $La_2(PDA)_3(H_2O)_5$ [121], which exhibited a blue emission at 408 nm attributed to the emissive organic PDA linkers. Compared with luminescence of the free ligand, $La_2(PDA)_3(H_2O)_5$ exhibits much enhanced blue light emission and slight red shifts, which was attributed to the enhancement of rigidity of the PDA ligands after coordinating to La^{3+} ions. Doping of a small amount of Tb^{3+} and Eu^{3+} led to the formation of isostructural $La_2(PDA)_3(H_2O)_5{:}Tb^{3+}$ and $La_2(PDA)_3(H_2O)_5{:}Eu^{3+}$, which can provide additional green (543 nm) and red (614 nm) emissions, respectively. Therefore, the white-light-emitting MOFs $La_2(PDA)_3(H_2O)_5{:}Tb^{3+}, Eu^{3+}$ can be realized by tuning the molar amount of Tb^{3+} and Eu^{3+} in the host framework solid. The optimized white-light-emitting MOFs, $La_2(PDA)_3(H_2O)_5{:}1.0\%Tb^{3+}, 2.0\%Eu^{3+}$ and $La_2(PDA)_3(H_2O)_5{:}1.5\%Tb^{3+}, 2.0\%Eu^{3+}$, exhibit the CIE coordinate of (0.3269, 0.3123) and (0.3109,0.3332), respectively.

Dang et al. prepared a family of lanthanide MOFs using a semirigid trivalent carboxylic acid as linker [208]. These MOFs are isostructural and display typical lanthanide luminescent emissions of Eu^{3+}, Tb^{3+}, Sm^{3+}, and Dy^{3+}, respectively. The absence of obvious emission from the ligand implies the efficient energy transfer from the semirigid ligands to Ln^{3+} ions. Furthermore, white-light emission was successfully obtained by codoping Dy/Eu and Dy/Sm into the Gd analogs. The CIE coordinates of $Dy_{0.02}Eu_{0.05}Gd_{0.93}L$ are (0.355, 0.313), which is very close to the ideal coordinate for pure white light (0.333, 0.333). Another sample by codoping Dy and Sm into Gd analog, $Dy_{0.01}Sm_{0.10}Gd_{0.89}L$, also exhibits the white-light emission with a CIE coordinates of (0.328, 0.320).

Recently, the pursuing of single white-light emitting materials to avoid the intrinsic color balance and device complication when using multiphosphors or multi-LEDs has attracted much attention. A direct white-light-emitting MOF, $[AgL]_n{\cdot}nH_2O$ (L = 4-cyanobenzoate), with tunable yellow to white luminescence by variation of excitation light was reported by Wang et al. [170]. The MOF displays a maximum emission at around 427 and 566 nm when excited at 355 and 330 nm, respectively. When irradiated at 330 nm, the emission intensity at 427 nm is decreased, but the emission intensity at approximately 513, 566, and 617 nm are enhanced, generating yellow luminescence. When adjusting the excitation light to 350 nm, the emission peaks at 427 and 566 nm are very strong which results in direct white light to the naked eye (Fig. 5). The CIE chromaticity coordinates of the white-light emissions excited at 350 and 349 nm are approximately (0.31, 0.33) and (0.33, 0.34), respectively, comparable to that of pure white light. Therefore, the MOF can be tuned from yellow to white by direct variation of excitation wavelength, indicating it can be potentially used as a single white phosphor for a white-light-emitting device equipped with a deep UV GaN LED, which has light output at 325–350 nm.

Wibowo et al. also presented two single component white-light-emitting phosphors based on the bismuth and lead MOFs [203]. When excited at 380 nm, the bismuth MOF $Bi_3(\mu_3\text{-}O)_2(pydc)_2(Hpydc)(H_2O)_2$ emits white composite light with a broad emission spectrum from about 400–600 nm, and three distinct maxima at 430, 460, and 480 nm with a shoulder around 556 nm can be identified, which is significantly blue shifted compared to the emission of ligand. This blue shift is attributed to LMCT transition and a change in the intraligand $\pi \rightarrow \pi^*$ and/or $n \rightarrow \pi^*$ transitions. Similarly, the lead MOF $Pb(pydc)(H_2O)$ exhibits slightly "whiter" composite photoluminescence, with a distinct emission maximum at 441 nm and three broad shoulders around 470, 520, and 563 nm. These studies have paved the way for the utilization of luminescent MOFs in white-light emitting devices.

Nenoff et al. reported the intrinsic broadband direct white-light emission originating from a single component MOF [202], which was synthesized based on indium ions, H_3BTB, and ox (BTB = 1,3,5-tris(4-carboxyphenyl)benzene, ox = oxalic acid), namely, $[In_3(BTB)_2(ox)_3]_n$. Interestingly, the material already emitted a white light owing to the broadband emission over the entire visible light region. In order to improve the intrinsic color properties such as color rendering index (CRI), correlated color temperature (CCT), and chromaticity to approach the requirements for solid-state lighting, the authors introduced a narrowband, red emission component into this system via Eu^{3+} doping. The simple mixing of Eu^{3+} metal source into the reaction solution of the indium framework successfully led to the formation of a

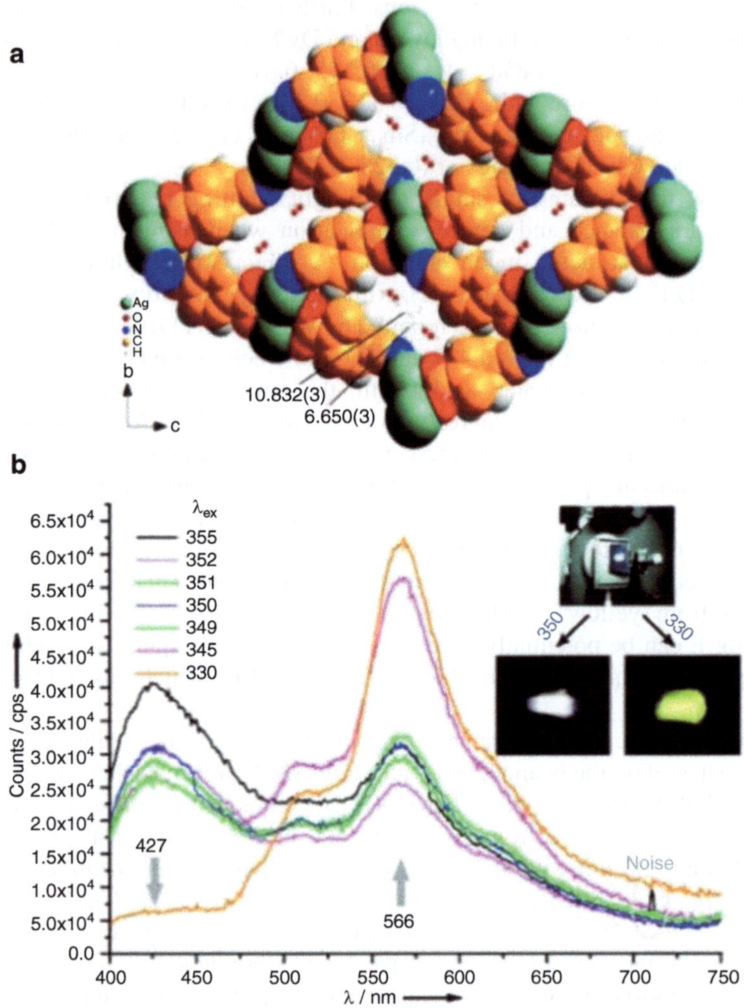

Fig. 5 (**a**) 3-D packing diagram of $[Ag(4\text{-cyanobenzoate})]_n{\cdot}nH_2O$ viewed along the a direction. (**b**) Solid-state PL spectra of $[Ag(4\text{-cyanobenzoate})]_n{\cdot}nH_2O$ by variation of excitation light under the same metrical condition. *Inset* shows the PL images of a sample excited by 350- and 330-nm light, respectively. Reprinted with the permission from [170]. Copyright 2009, American Chemical Society

mixed-metal framework. By increasing the Eu^{3+} concentration to 10%, the CRI and CCT shifted closer to the set target of CRI ~90 and CCT ~3,200 K. When the sample is excited at 350, 360, 380, and 394 nm, the coordinates are (0.369, 0.301), (0.309, 0.298), (0.285, 0.309), and (0.304, 0.343), respectively, which closely approach targeted values (Fig. 6). The absolute quantum yield of the mixed MOF was found to be 4.3% when excited at 330 nm, which is modest as compared to traditional phosphors, but falls within the expected range for similar reported systems. This

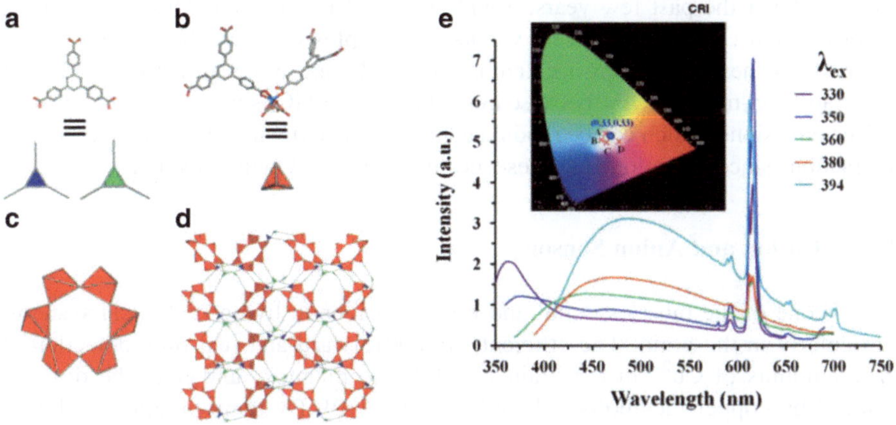

Fig. 6 Molecular building blocks in MOF $[In_3(BTB)_2(ox)_3]_n$, corresponding to (**a**) two topologically distinct 3-connected nodes and (**b**) a 4-connected node; (**c**) corner-sharing tetrahedra forming a 6-membered ring; and (**d**) topological representation of single net in $[In_3(BTB)_2(ox)_3]_n$; (**e**) emission spectra of 10% Eu-doped $[In_3(BTB)_2(ox)_3]_n$ when excited between 330 and 380 nm; inset: 1931 CIE chromaticity diagram highlighting corresponding chromaticity coordinates (*A–D*) approaching targeted values. Reprinted with the permission from [202]. Copyright 2012, American Chemical Society

result provides a new path for the rational design of alternative materials for white-light emission.

4.3 Chemical Sensors

Of all the MOF applications, chemical sensor is clearly a promising application due to the luminescent properties of MOFs are very sensitive to and dependent on their structural characteristics, coordination environment of metal ions, nature of the pore surfaces, and their interactions with guest species through coordination bonds and π–π interactions and hydrogen bonding. By the incorporation of analyte into their pores, the MOFs can exhibit different degrees of luminescence enhancement or quenching behaviors, which make the MOF materials ideal for molecular recognition. The permanent porosity within some porous luminescent MOFs has enabled the reversible uptake and release of some sensing substrates; thus the exploration of reversible luminescent sensing MOFs will be feasible, so sensing MOFs can be regenerated and repeatedly utilized. The high surface areas make the MOFs serve as a preconcentrator to concentrate analytes to levels high above those in the surrounding atmosphere, thus leading to much smaller detection levels than otherwise available. Moreover, the tunable pore sizes and functional sites such as Lewis basic/acidic sites and open metal sites afford unprecedented selectivity either through size exclusion or chemical interactions, while mesoporous nature of some mesoporous luminescent MOFs will make sensing of some large molecules such as biologically active species possible. Over the past few years, a wide range of luminescent MOFs for sensing cations, anions, small molecules, vapors, and explosives have been realized and reported. In these MOF sensors, lanthanide MOFs are more attractive candidates than transition-metal MOFs, because many lanthanide MOFs have sharp and characteristic emissions, which can be modulated by both analyte–metal and analyte–ligand interactions since the unique luminescence mechanism of lanthanide ions.

4.3.1 Cation and Anion Sensor

Cu^{2+} is one of the most essential and important ions in living biological systems, particularly in the brain. The efficient, straightforward and real-time detection of trace amounts of Cu^{2+} in the brain is helpful to diagnose and treat the diseases induced by copper metabolism disorders [209, 210]. By immobilizing the Lewis basic sites within porous MOFs as sensing sites, Chen et al. develop a new way of sensing metal ions. These types of MOFs are generally difficult to prepare because such Lewis basic sites tend to bind other metal ions to form condensed structures. Exploiting the preferential binding of lanthanide ions to carboxylate oxygen atoms over pyridyl nitrogen atoms in Ln^{3+}–pyridinecarboxylate MOF, they synthesized a luminescent MOF $Eu(PDC)_{1.5}(DMF)\cdot(DMF)_{0.5}(H_2O)_{0.5}$ (PDC = pyridine-3, 5-dicarboxylate) [131]. The MOF is porous, and the basic Lewis sites line the interior of the channels, which may be emptied of DMF and water molecules by heating at 200°C. Due to the different binding of these free Lewis basic pyridyl sites with metal ions, the activated MOF $Eu(PDC)_{1.5}$ exhibits a sensing function for metal ions: alkali metal ions and alkaline-earth metal ions have a negligible effect on the luminescence intensity, whereas Co^{2+}, especially Cu^{2+}, can reduce the luminescence intensity significantly (Fig. 7). It is speculated that the binding of the pyridyl nitrogen atoms to Cu^{2+} or Co^{2+} reduces the antenna efficiency of the PDC organic linkers to magnify the f–f transitions of Eu^{3+}, resulting in quenching of the luminescence. These results indicate that the Lewis basic sites within porous MOFs are expected to play very important roles for their recognition of small Lewis acidic molecules and metal ions.

Qian and Chen et al. also demonstrated the first example of Cu^{2+} sensing in aqueous solution and simulated physiological aqueous solution using lanthanide MOFs [132]. They prepared a luminescent MOF named as $Eu_2(FMA)_2(OX)(H_2O)_4\cdot 4H_2O$ (FMA = fumarate; ox = oxalate), whose luminescent intensity is heavily metal ion dependent: alkaline metal ion and alkaline-earth metal ions have basically no effect on the luminescence intensity, while Cu^{2+} has the largest quenching effect, and the fluorescence lifetime of 394.60 ms in $Eu_2(FMA)_2(OX)(H_2O)_4\cdot 4H_2O$ is very significantly reduced to 30.45 ms in the presence of 10^{-2} M Cu^{2+}. The sensing of Cu^{2+} in the simulated physiological aqueous solution is also explored, and the comparable result is observed, indicating this MOF is suitable to probe Cu^{2+} ion in a biological system. A similar example for sensing of Cu^{2+} and Co^{2+} in aqueous solution is reported by Luo et al. [116]. Instead of lanthanide MOFs, they utilized a lanthanide-doped approach to obtain luminescent MOF. The Eu^{3+}- and Tb^{3+}-doped MOFs were prepared by immersing the $[NH_4]_2[ZnL]\cdot 6H_2O$

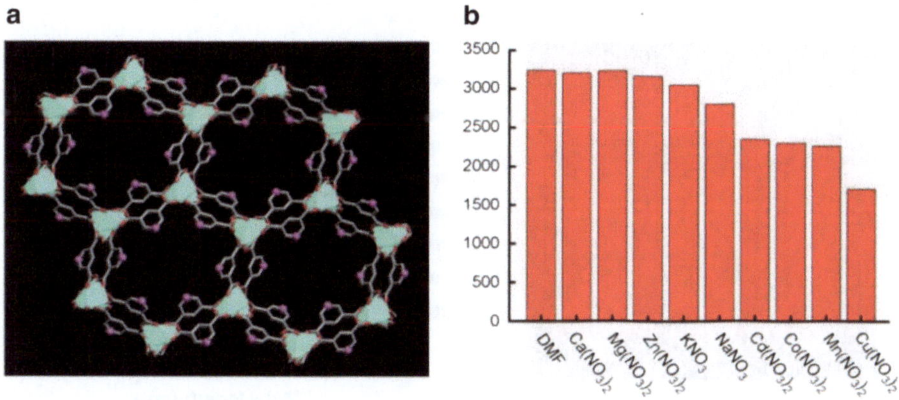

Fig. 7 (**a**) Crystal structure of $Eu(PDC)_{1.5}(DMF)\cdot(DMF)_{0.5}(H_2O)_{0.5}$, viewed along *a* axis indicating immobilized Lewis basic pyridyl sites oriented toward pore centers. C gray, N purple, O red, Eu green polyhedra. Hydrogen atoms, terminal DMF molecules, and solvent molecules are omitted for clarity. (**b**) Comparison of the luminescence intensity of dehydrated MOF $Eu(PDC)_{1.5}$ incorporating different metal ions in 10-mM DMF solutions of $M(NO_3)_x$. Reprinted with the permission from [131], Copyright 2009, John Wiley & Sons Ltd.

(L = 1,2,4,5-benzenetetracarboxylate) in $EuCl_3$ or $Tb(ClO_4)_3$ solutions through ion exchange of NH_4^+. Subsequently, the obtained lanthanide-doped MOFs were immersed in the MCl_x (M = Na^+, K^+, Zn^{2+}, Ni^{2+}, Mn^{2+}, Co^{2+}, Cu^{2+}) solutions to monitor the luminescent intensity. These doped MOFs show a negligible effect on the luminescence intensity for the Na^+, K^+, and Zn^{2+}, while they are highly sensitive to the Co^{2+} (in Tb@MOF) and Cu^{2+} (in Eu@MOF), respectively.

The sensing of anions is also an interesting theme because many inorganic anions play fundamental roles in environmental and biological systems. Chen et al. reported a novel luminescent MOF, Tb(BTC)·G (MOF-76, BTC = benzene-1,3,5-tricarboxylate, G = guest solvent) with OH groups in the terminal solvents instead of in the organic linkers [143]. MOF-76 exhibits increased luminescence upon incorporated with Br^-, Cl^-, F^-, SO_4^{2-}, and CO_3^{2-} in methanol. Because F^- has much stronger interactions with the terminal methanol molecules to confine the O–H bond stretching and thus reduce its quenching effect, the addition of F^- enhances the luminescence intensity by as much as a factor of four over the non-F^--incorporated MOF-76 (Fig. 8). This selective sensing function with respect to F^- is also observed in the solvent of DMF, featuring the bright promise for these kinds of porous luminescent MOFs for the sensing of fluoride anion.

In most examples presented above, the sensing function is mainly based on lanthanide MOFs; a particular work which relies on ligand-centered fluorescence for sensing of anions is provided by Qiu et al. [211]. In their report, a highly symmetric tetragonal cadmium-based MOF with 5-methyl-1H-tetrazole as ligand was synthesized. In DMF suspension, this MOF exhibits an emission at around 370 nm attributed to the intraligand transitions, whose intensity decreased continuously upon adding NO_2^-, while there was only a negligible effect on the luminescence intensity for

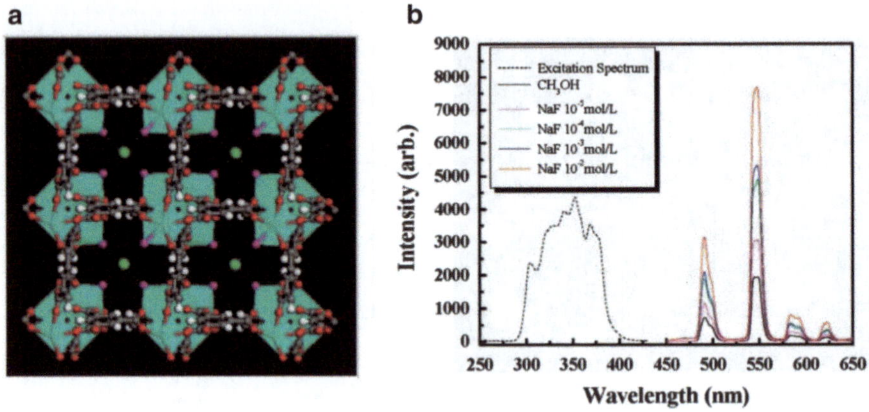

Fig. 8 (**a**) Single crystal X-ray structure of MOF-76b activated in methanol containing NaF with the model of fluoride (*green*) at the center of the channel involving its hydrogen-bonding interaction with terminal methanol molecules (methanol oxygen, purple; the methyl group from methanol is omitted for clarity). (**b**) Excitation (*dotted*) and PL spectra (*solid*) of MOF-76b solid activated in different concentrations of NaF methanol solution (excited and monitored at 353 and 548 nm, respectively). Reprinted with the permission from [143]. Copyright 2008, American Chemical Society

ClO_4^-, NO_3^-, Cl^-, and I^- and the same luminescence response was also observed in a water suspension.

4.3.2 Molecule Sensor

Recently, Chen et al. demonstrated that the open luminescent lanthanide sites such as Eu^{3+} and Tb^{3+} will play a crucial role in molecular recognition processes as binding of substrates [133]. They prepared a MOF $Eu(BTC)(H_2O)\cdot 1.5H_2O$ (BTC = benzenetricarboxylate), which is isostructural with MOF-76 and exists with 1D channels of about 6.6 × 6.6 Å (Fig. 9). The luminescence intensity of activated EuBTC is largely dependent on the solvent molecules, particularly in the case of DMF and acetone, which exhibit the most significant enhancing and quenching effects, respectively. Furthermore, a gradual decrease of the fluorescence intensity was observed upon the addition of acetone to the 1-propanol emulsion of EuBTC, and the $^5D_0 \rightarrow {}^7F_2$ emission intensity of Eu^{3+} versus the volume ratio of acetone could be well fitted with a first-order exponential decay. It is suggested that the binding interaction of the open metal sites with guest solvent molecules definitely plays an important role, and the weakly coordinated 1-propanol molecules on the Eu^{3+} sites can be gradually replaced by DMF and acetone molecules, leading to the luminescence enhancement and diminishment, respectively.

In addition, Chen et al. prepared a NIR luminescent ytterbium MOF Yb(BPT) $(H_2O)\cdot(DMF)_{1.5}(H_2O)_{1.25}$ (BPT = biphenyl-3, 4′, 5-tricarboxylate) for sensing of small molecules [149]. This MOF crystallizes in a tetragonal space group $P4_3$. Each ytterbium atom is coordinated by six oxygen atoms from the carboxylate groups of

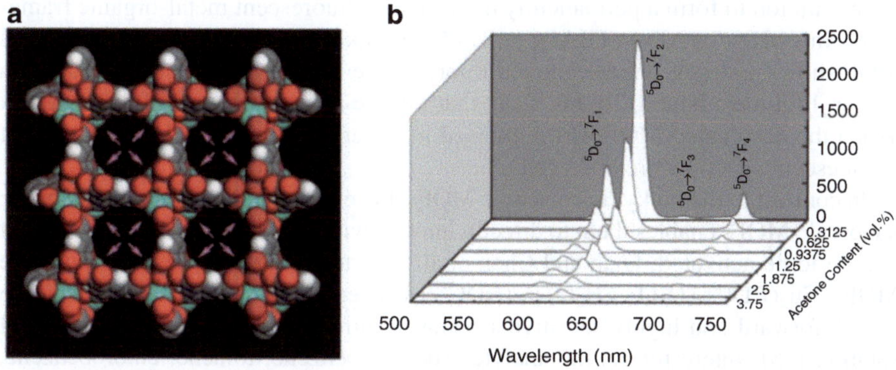

Fig. 9 (**a**) X-ray crystal structure of $Eu(BTC)(H_2O)\cdot 1.5H_2O$ viewed along the *c* axis, exhibiting 1D channels of about 6.6 × 6.6 Å, and uniformly immobilized accessible Eu^{3+} sites within the framework, shown by the *pink arrows*. The free and terminal water molecules are omitted for clarity; Eu, *light green*; O, *red*; C, *gray*; H, *white*. (**b**) The PL spectra of 1-propanol emulsion of EuBTC in the presence of various content of acetone solvent. Reprinted with the permission from [133], Copyright 2007, John Wiley & Sons Ltd

BPT and one terminal water molecule. Yb atoms are bridged by BPT organic linkers to form a 3D rod-packing structure. The activated MOF Yb(BPT) exhibits typical NIR emission from the $^2F_{5/2} \rightarrow {^2F_{7/2}}$ transition of the Yb^{3+} ion at 980 nm, when excited at 326 nm. Similar to the EuBTC, the luminescent intensity exhibit the most significant enhancing and quenching effects in the case of DMF and acetone, respectively. The decreasing trend of the fluorescence intensity of the $^2F_{5/2} \rightarrow {^2F_{7/2}}$ transition of Yb^{3+} at 980 nm versus the volume ratio of acetone could be well fitted with a first-order exponential decay, indicating that fluorescence quenching of Yb(BPT) by acetone is diffusion controlled. This work extends the sensing function of luminescent MOFs into the NIR region, providing their potentials for the sensing of substrates in biological systems.

Another example of recognition of small solvent molecules was reported by Lin et al. [145], who synthesized a new family ytterbium–organic frameworks with 4,4′-oxybis(benzoate) (OBA) ligands and suitable cationic species. The MOF Na $[Tb(OBA)_2]_3\cdot 0.4DMF_3\cdot 1.5H_2O$ exhibits the strongest emission intensities in BuOH and EtOH suspensions and significantly decrease in MeOH and H_2O suspensions gradually increasing luminescence intensities when dispersed sequentially in water, methanol, and ethanol as suspensions. Unlike the sensing result of Eu(BTC) mentioned above, this MOF does not show its strongest emission in a DMF suspension, which was suspected that the different sizes and geometries of active metal centers were exposed in these frameworks, such as six- and eight-coordinated metal sites in the desolvated Eu(BTC) and $Na[Tb(OBA)]_2$ solids, respectively, thus resulting in their own unique sensing abilities.

An example for recognition of small solvent molecules based on ligand-centered luminescence was performed in an In–pyrenetetrabenzoic acid MOF, reported by Stylianou et al. [201]. They selected a highly fluorescent pyrene-derived ligand with an indium ion to form a permanently microporous fluorescent metal-organic framework, $In_2(OH)_2$(TBAPy) (TBAPy = 1,3,6,8-tetrakis(*p*-benzoic acid)pyrene). This MOF shows a strong linker-centered fluorescence emission band at 471 nm and a longer lifetime about 0.11 ms than that of the uncoordinated TBAPy ligand. When the desolvated MOF was immersed in different solvents, solvent-dependent luminescent spectra were observed.

In contrast to the bulky macroscopic MOFs, the excellent dispersible nature of the nanoscale MOFs enables them to directly interact with chemical species and does not require to be activated. Qian and Chen et al. reported a rare luminescent nanoscale MOF $Eu_2(BDC)_3(H_2O)_2{\cdot}(H_2O)_2$ (BDC = benzene-1,4-dicarboxylate), for the straightforward and highly sensitive sensing of nitroaromatic explosives in ethanol solution [75]. Interestingly, the analytes such as benzene, toluene, chlorobenzene, phenol, *o*-cresol, and 4-bromophenol basically do not affect the luminescence intensity, while the nitroaromatic compounds such as nitrobenzene, 2,4-DNT (2,4-dinitrotoluene), and TNT (2,4,6-trinitrotoluene) significantly quench the luminescence of $Eu_2(BDC)_3(H_2O)_2{\cdot}(H_2O)_2$ in ethanol, which is attributed to a competition of absorption of the light source energy and the electronic interaction between the nitroaromatic compounds and BDC ligands. The variety of the organic ligands which can be assembled into the nanoscale MOFs for their matching competitive absorption with the different analytes highlights this strategy as a very promising approach for the sensing of small molecules.

Detection of *Bacillus anthracis* is an essential process in the event of biological warfare since these bacterial endospores serve as a delivery vehicle in anthrax attacks. The luminescent detection of bacterial endospores is based on the presence of DPA (DPA = dipicolinic acid) in the endospore casing, which weighs up to 15% of the endospores. Qian and Chen described the highly sensitive, selective, and instant turn-on sensing of bacterial endospores using a nanoscale MOF, $Eu_2(FMA)_2(OX)(H_2O)_4{\cdot}4H_2O$ (FMA = fumarate, ox = oxalate) [212]. Interestingly, the sizes and morphologies of the nanoscale MOF can be simply tuned and controlled by the addition of different amounts of the CTAB surfactant (CTAB = cetyltrimethylammonium bromide), matching with the simulatation based on the well-established Bravais–Friedel–Donnay–Harker (BFDH) method very well. The nanoscale nature of $Eu_2(FMA)_2(OX)(H_2O)_4{\cdot}4H_2O$ has not only allowed the easily disperse in solvent, but also significantly enhanced the sensitivity level up to about 90 times for 2 ppm DPA sensing. Such highly sensitive sensing of DPA is basically not affected by other components such as isophthalic acid, phthalic acid, terephthalic acid, 1,3,5-benzenecarboxylic acid, D-phenylalanine, sodium benzoate, riboflavin, DL-tryptophan, and different ions which coexist with DPA in the bacterial endospores. This result suggests very bright promise for developing controllable nanoscale functional MOF sensors.

The luminescent MOF sensors mentioned above generally detect a single target molecule among several molecules, but cannot differentiate multiple molecules simultaneously. Recently, a clever molecular decoding strategy using MOFs as hosts has been developed by Takashima et al. [213]. Amazingly, the MOFs can accommodate a class of molecules and further distinguish among them with characteristic different visible light emission for each different guest molecule. They synthesized a 1,4,5,8-naphthalenediimide (NDI)-based interpenetrating MOF, $Zn_2(1,4\text{-}BDC)_2(dpNDI)\cdot 4DMF$ (1,4-BDC = 1,4-benzenedicarboxylate; dpNDI, *N,N′*-di(4-pyridyl)-1,4,5,8-naphthalenediimide), which will undergo a dynamic structural transformation to confine a class of aromatic volatile organic compounds or tropospheric air pollutants. In the structure of the interpenetrating MOF $Zn_2(1,4\text{-}BDC)_2(dpNDI)\cdot 4DMF$, one of the frameworks is intergrown at the central position of the void that is created by the other framework. The void spaces can be classified as two different environments: the characteristic slit between dpNDI and 1,4-BDC (site A) and the remaining void space (site B). The aromatic volatile organic compounds (VOCs) such as benzene, toluene, xylene, anisole, and iodobenzene can be introduced by immersion of a dry MOF into the liquid of each aromatic VOC. The guest-exchanged MOF shows a significant framework displacement, in which one framework slid along the main axis of one 1,4-BDC molecule. The MOF $Zn_2(1,4\text{-}BDC)_2(dpNDI)$ shows a very weak fluorescence with a very low quantum yield and a short average lifetime. In contrast, all VOC-incorporated $Zn_2(1,4\text{-}BDC)_2(dpNDI)$ exhibit intense fluorescence in the visible light region, and the emission color is dependent on the chemical substituent of the aromatic guest. Furthermore, the normalized emission spectra exhibit a gradual red shift as the electron-donating capability of VOCs increased upon excitation at 370 nm (Fig. 10). The multicolor luminescence is attributed to the enhanced naphthalenediimide-aromatic guest interaction through the induced-fit structural transformation of the interpenetrating framework.

4.3.3 Gas and Vapor Sensor

Gas sensors have a wide range of application in the fields of aerodynamics, environmental analysis, analytical chemistry, and biochemistry [214, 215]. Similar to the sensing of small molecules and ions, the luminescent properties of MOFs can also be perturbed by gases, thereby providing an effective means for sensing of gases. Cyclometalated iridium complexes such as $Ir(ppy)_3$ (ppy = 2-phenylpyridine) are highly efficient phosphorescent molecules and can be used to accurately detect oxygen due to their MLCT phosphorescence and can be readily quenched by molecules with a triplet ground state. Xie et al. recently synthesized a Zn–Ir (2-phenylpyridine) MOF using $Ir(ppy)_3$ derivatives [200], which possesses open channels of 7.9 × 4.3 Å that are perpendicular to the (1, −1, 6) plane and contain guest solvent molecules. To demonstrate the applicability for O_2 sensing, the MOF was gradually dosing in O_2 from 0.05 to 1.0 atm, and the gradual decreases in luminescence intensity at 538 nm were observed. Furthermore, the luminescence measurements after alternating cycles of O_2 dosing at 0.1 atm (30 s) and O_2 removal under vacuum (120 s) indicate that the luminescence is reversibly quenched by O_2, with $<5\%$ of the original luminescence lost after eight cycles.

Song et al. reported a lanthanide MOF, $[Eu_2L_3(H_2O)_4]\cdot 3DMF$ (L = 2′,5′-bis (methoxymethyl)-[1,1′:4′,1″-terphenyl]-4,4″-dicarboxylate), for sensing of DMF

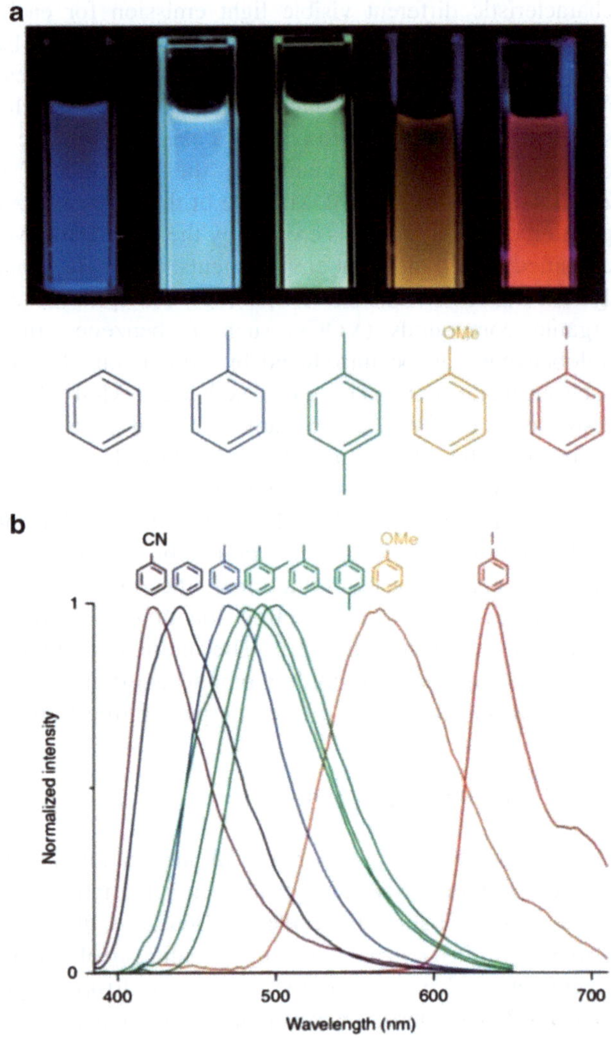

Fig. 10 Multicolor luminescence of VOC incorporated $Zn_2(bdc)_2(dpNDI)$. (**a**) The resulting luminescence of crystal powders of $Zn_2(bdc)_2(dpNDI)$ suspended in each VOC liquid after excitation at 365 nm using a commercial ultraviolet lamp. (**b**) Height-normalized luminescent spectra of VOC incorporated $Zn_2(bdc)_2(dpNDI)$ after excitation at 370 nm. Reprinted with the permission from [213]. Copyright 2011, Macmillan Publishers Ltd

vapor [216]. The water-exchanged MOF displays a much weaker luminescence under UV irradiation, while exhibits more than eightfold enhancement of luminescence after incubation under DMF vapor (Fig. 11). This DMF-triggered turn-on luminescence was rationalized by the DMF–ligand interactions that presumably shift the excited state energy level of the ligand and thus facilitates the ligand–lanthanide energy transfer process. The response rate of this MOF sensor is quite fast, with 95% of turn-on and turn-off achieved within a few minutes and 10–20 s, respectively. This example offers the potential to build lanthanide-based turn-on luminescent MOF sensors by proper ligand design, targeting analytes through ligand–analyte interactions.

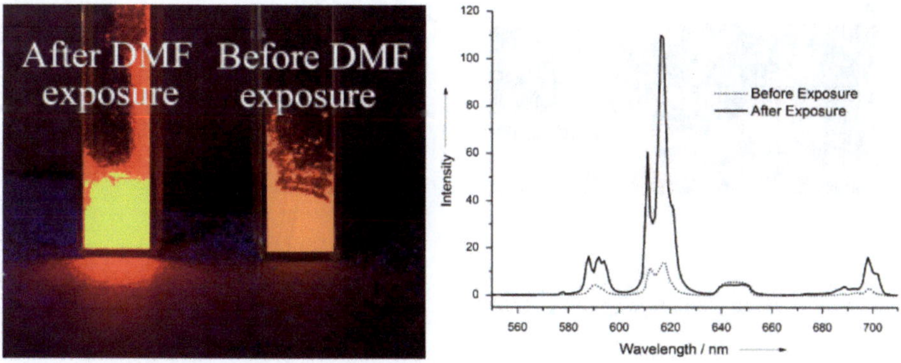

Fig. 11 Photograph (*left*) and emission spectra (*right*) of $Eu_2L_3(H_2O)_4$ (L = 2′,5′-bis (methoxymethyl)-[1,1′:4′,1″-terphenyl]-4,4″-dicarboxylate) before and after exposure to DMF vapor. Reprinted with the permission from [216], Copyright 2013, John Wiley & Sons Ltd

Chemical sensors for rapid detection of explosives in the gas phase are attracting increasing attention owing to homeland security, environment monitoring, and humanitarian implications. Lan et al. reported a MOFs $Zn_2(bpdc)_2(bpee)$ (bpdc = 4,4′-biphenyldicarboxylate; bpee = 1,2-bipyridylethene) for detection of both DNT and DMNB (2,3-dimethyl-2,3-dinitrobutane) with rapid response and high sensitivity [157]. Within ten seconds, the fluorescence quench percentages of the MOFs thin film reach almost the maxima for both DNT and DMNB (ca. 85% and 84%, respectively). The sensitivity of the film for DMNB exceeds previously reported materials based on conjugated polymer thin films. This was attributed to the infinite 3D framework structure, the large surface area, and the pore confinement of the analyte inside the molecular-sized cavities which facilitates stronger interactions between the DMNB and the framework. Considering the very fast responses and high sensitivity, it would be possible to produce a new and important sensor for explosives using the MOFs.

4.3.4 Ratiometric Sensor

In most examples presented above, the sensing function is based on changes in the intensity of one transition, which can be heavily affected by the quantity of the luminophore, excitation power, and the drifts of the optoelectronic system. Thus, the comparison of the emission of different samples based on the detected intensity may lead to erroneous conclusions. Although the measurements of quantum yields and/or lifetime are affected neither by the intensity of the excitation source nor by the probe concentration, they require a relatively long time and the computational treatment. The utilizing of the ratio between the intensity of two transitions of the same luminescent material, instead of only one transition, can overcome the main drawbacks of the intensity-based measurements of only one transition. The ratiometric

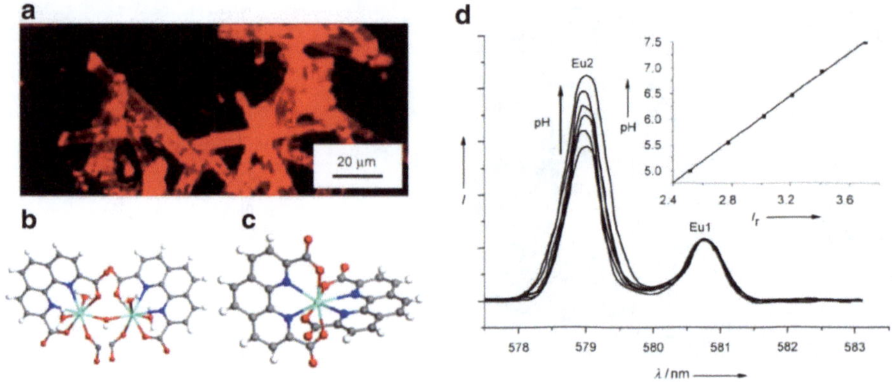

Fig. 12 (**a**) Optical microscopy image of ITQMOF-3-Eu under UV light. (**b**) Eu2 coordination environment. (**c**) Eu1 coordination environment. C, *gray*; H, *white–gray*; N, *blue*; Eu, *green*; O, *red*. (**d**) Intensity variation of the Eu2 $^5D_0 \rightarrow {}^7F_0$ transition from pH value of 7.5 to 5; the inset shows the linear variation of intensity ration of the $^5D_0 \rightarrow {}^7F_0$ emissions with the pH value. Reprinted with the permission from [138], Copyright 2009, John Wiley & Sons Ltd

luminescent sensor enables the sensing independent of the sample concentration and of the drifts of the optoelectronic system such as lamp and detectors.

Harbuzaru et al. have given a detailed description of the use of the Eu^{3+}-MOF ITQMOF-3-Eu as ratiometric pH sensors. The MOF ITQMOF-3-Eu was synthesized using 1,10-phenanthroline-2,9-dicarboxylic acid (H_2PhenDCA) as ligand, in which both carboxylate and phenanthroline moieties may coordinate to the metal center [138]. The MOF ITQMOF-3-Eu can be described as a layered MOF with two well-defined sheets, and each sheet contains one type of Eu^{3+} ions (Eu1 or Eu2) in a different crystallographic position. Due to the two different Eu^{3+} environments, ITQMOF-3-Eu exhibits two nondegenerate $^5D_0 \rightarrow {}^7F_0$ emission lines at 579.0 (Eu2) and 580.7 (Eu1) nm, respectively (Figure 12). The luminescence features of each Eu^{3+} site may be rationalized in terms of the relationship between the covalency of the Eu–(O,N) bonds and the nephelauxetic effect that influences the energy of the $^5D_0 \rightarrow {}^7F_0$ transition: a site with more covalent Eu-(O,N) bonds has a $^5D_0 \rightarrow {}^7F_0$ transition at lower energy and a longer emission lifetime. Intriguingly, only one of the two Eu^{3+} emitting sites is affected by the pH variation in the range 5–7.5; thus the sensor does not require external calibration. By following the intensity ratio of the $^5D_0 \rightarrow {}^7F_0$ emissions of the two Eu^{3+} types, it is possible to determine the pH of the solution. The high-emission quantum efficiency and pH-sensing capability enable ITQMOF-3-Eu to be used in a new miniaturized pH sensor prototype by combining the material with a commercial fiber optic 1.5 mm in diameter.

Recently, Qian and Chen et al. reported the first ratiometric thermometer based on luminescent MOF, $Eu_{0.0069}Tb_{0.9931}$-DMBDC (DMBDC = 2,5-dimethoxy-1,4-benzenedicarboxylate) [198]. The advantages of using luminescence over other conventional thermometers are fast response, high sensitivity, noninvasive operation,

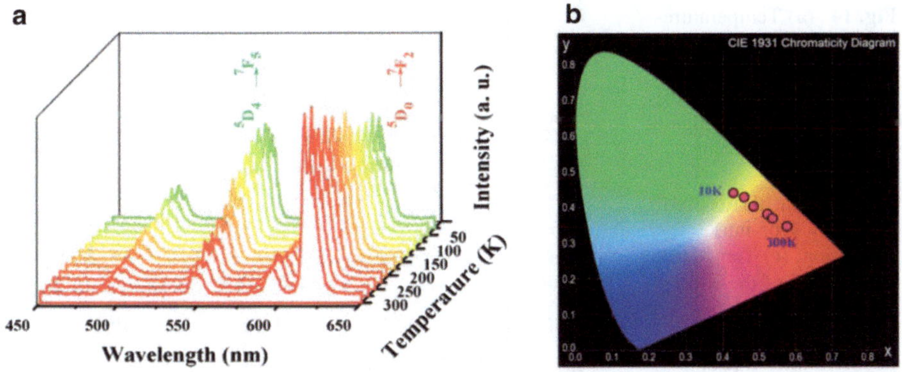

Fig. 13 (**a**) Emission spectra of $Eu_{0.0069}Tb_{0.9931}$-DMBDC recorded between 10 and 300 K; (**b**) CIE chromaticity diagram showing the luminescence color of $Eu_{0.0069}Tb_{0.9931}$-DMBDC at different temperatures. Reprinted with the permission from [198]. Copyright 2012, American Chemical Society

and inertness to strong electric and magnetic fields. To date, most luminescence-based thermometers rely on a single emission, whose accuracy can be heavily affected by the quantity of the luminescent materials and excitation power. To overcome this issue, the luminescent intensity ratio I_{Tb}/I_{Eu} of the mixed MOF instead of intensity was utilized to realize the temperature sensing. Generally, the luminescent intensity of Tb-DMBDC and Eu-DMBDC decreases gradually as the temperature increases due to the thermal activation of nonradiative-decay pathways. However, the mixed Tb/Eu MOF $Eu_{0.0069}Tb_{0.9931}$-DMBDC exhibits a significantly different temperature-dependent luminescent behavior from those of Tb-DMBDC and Eu-DMBDC. The emission intensity of the Tb^{3+} ions in $Eu_{0.0069}Tb_{0.9931}$-DMBDC decreases, while that of the Eu^{3+} increases with the temperature (Fig. 13). The main emission bands of 613 (Eu^{3+}) and 545 (Tb^{3+}) nm were comparable at 10 K, while the emission of Eu^{3+} almost dominates the whole spectrum at 300 K. This is attributed to the phonon-assisted Förster energy transfer, which results in an efficient energy transfer from Tb^{3+} to Eu^{3+} at high temperatures, as evidenced by luminescent lifetime measurements. In addition, the tunable luminescence colors from the green yellow to red from 10 to 300 K allow directly visualizing the temperature change instantly and straightforwardly. Such a Tb/Eu MOF featuring temperature-dependent luminescence enables itself to be an excellent candidate for self-referencing luminescent thermometers as no further calibration of luminescence intensity is required.

The similar ratiometric temperature sensing was demonstrated by Hasegawa et al. using a mixed MOF, $[Tb_{0.99}Eu_{0.01}(hfa)_3(dpbp)]_n$ (hfa = hexafluoro acetylacetonato, dpbp = 4,4′-bis(diphenylphosphoryl) biphenyl) in the range of 200 to 500 K [217]. The MOF exhibits the temperature-dependent emission. The emission intensities at 543 nm of $[Tb_{0.99}Eu_{0.01}(hfa)_3(dpbp)]_n$ decrease dramatically with increasing temperature. In contrast, the emission intensities at 613 nm increase

Fig. 14 (**a**) Temperature-dependent emission spectra of mixed MOF $[Tb_{0.99}Eu_{0.01}(hfa)_3(dpbp)]_n$ in the temperature range of 200–450 K; (**b**) Color pictures of $[Tb_{0.99}Eu_{0.01}(hfa)_3(dpbp)]_n$ under UV (365 nm) irradiation; (**c**) Temperature dependence of the energy transfer efficiency ($\eta_{Tb\text{-}Eu}$). Reprinted with the permission from [217], Copyright 2013, John Wiley & Sons Ltd

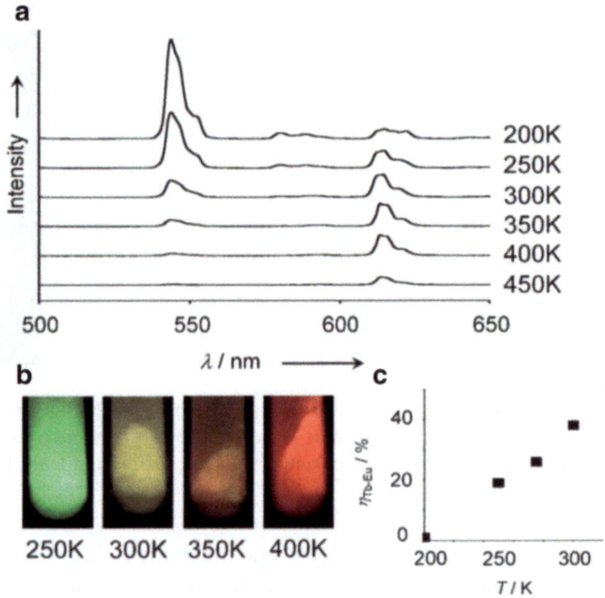

slightly (Fig. 14). This MOF exhibits brilliant green, yellow, orange, and red photoluminescence under UV irradiation at 250, 300, 350, and 400 K, respectively.

Rocha and Carlos et al. also demonstrated the nanoparticles of MOF $Tb_{0.99}Eu_{0.01}(BDC)_{1.5}(H_2O)_2$ (BDC = 1-4-benzendicarboxylate) as ratiometric luminescent nanothermometers in the physiological temperature (300–320 K) range [218]. The nanoparticles were prepared by a reverse microemulsion method with average length and diameter of 300 and 30 nm, respectively. Aqueous suspensions of the nano-MOF display an emission quantum yield of 0.23 excitation at 320 nm and relative sensitivity of 0.37% K^{-1} at 318 K, suggesting the possibility of using nano-MOFs to measure physiological temperatures. These MOF nanoparticles present a considerable potential for applications of biological interest, particularly in multimodal such as thermal, optical, and magnetic resonance imaging.

4.4 Application in Biomedicine

Recently, the use of nanoscale lanthanide MOFs for biological and biomedical applications has attracted increasing attentions. The interest of these functional materials relies on the combination of the chemical or biofunctional behavior of MOFs and the unique luminescence properties of lanthanide ions, such as high photostability, long decay rates, large Stokes shifts, and narrow emission bands. Besides their luminescent characteristics, lanthanide MOFs can possess paramagnetic properties which help to increase the relaxation rate of water protons in the

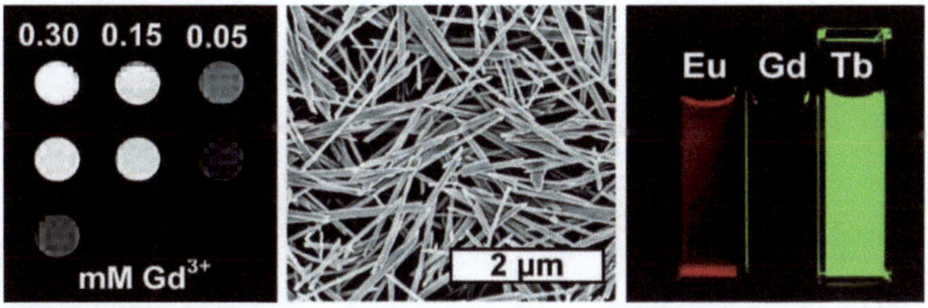

Fig. 15 MR images of suspensions of $Gd_{0.95}(BDC)_{1.5}(H_2O)_2$ in water containing 0.1% xanthan gum and luminescence images of ethanolic suspensions of $Gd_{0.95}(BDC)_{1.5}(H_2O)_2$, $Gd_{0.95}(BDC)_{1.5}(H_2O)_2$:$Eu_{0.05}$, and $Gd_{0.95}(BDC)_{1.5}(H_2O)_2$:$Tb_{0.05}$. Reprinted with the permission from ref. [73]. Copyright 2006, American Chemical Society

tissues being imaged, making them useful as contrast agents in magnetic resonance imaging (MRI) spectroscopy.

Multimodal imaging is a new imaging technique which combines more than one imaging modality, such as X-ray, nuclear, ultrasound, computed tomography (CT), MRI, and fluorescence imaging. Multimodal imaging is becoming more popular because of its improved sensitivity, high resolution, and morphological visualization. In particular, the combination of fluorescence imaging and MRI can ally the sensitivity of the fluorescence component with the high degree of spatial resolution of MRI. Lin et al. synthesized nanorods of MOF $Gd(BDC)_{1.5}(H_2O)_2$ through a reverse microemulsion method, which allows the control of the morphologies and sizes by alteration of the water-surfactant ratio of the microemulsion system [73]. These nanomaterials display large longitudinal relaxivities ($R1$) of 35.8 $mM^{-1}\ s^{-1}$ and transverse relaxivities ($R2$) of 55.6 $mM^{-1}\ s^{-1}$ on a per Gd^{3+} basis and extraordinarily large $R1$ of $1.6 \times 10^7\ mM^{-1}\ s^{-1}$ and $R2$ of $2.5 \times 10^7\ mM^{-1}\ s^{-1}$ on a per nanoparticle basis. The level of $R1$ is unprecedented and at least an order of magnitude higher than those of Gd^{3+}-containing liposomes which have been shown to be effective target-specific MRI contrast agents for cancer and cardiovascular disease. In addition, the analogs $Gd_{0.95}(BDC)_{1.5}(H_2O)_2$:$Eu_{0.05}$ and $Gd_{0.95}(BDC)_{1.5}(H_2O)_2$:$Tb_{0.05}$ were also synthesized. Ethanol suspensions of these materials are highly luminescent upon UV excitation with characteristic red and green luminescence from Eu^{3+} and Tb^{3+}, respectively, suggesting that they can be used as potential contrast agents for multimodal imaging (Fig. 15).

The combination of molecular imaging and drug delivery in MOFs will lead to the exciting possibility of MOF-based theranostics, in which imaging will be used to guide, follow, and quantify the drug delivery process. Up to now, core–shell architectures combining diverse functionalities and surface modifications into a single nano-MOF have been designed as biosensing platforms for imaging, targeting, diagnostics, and therapy. Lin et al. demonstrated a novel strategy of delivering a fluorescence-imaging contrast agent and an anticancer drug by postsynthetic modifications of a highly porous MOF [87]. They synthesized an amino-functionalized iron–carboxylate MOF by incorporating 2-aminoterephthalic

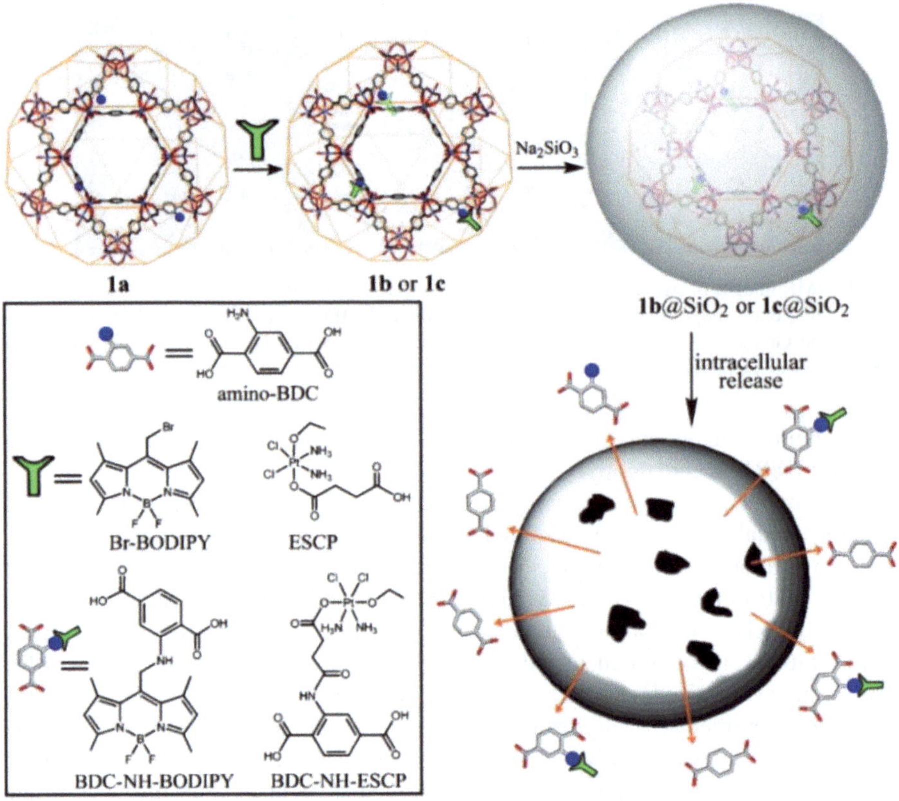

Fig. 16 Functionalization of iron–carboxylate MOF with an optical imaging contrast agent (BODIPY) and a prodrug of cisplatin (ESCP) through postsynthetic modifications. Reprinted with the permission from [87]. Copyright 2009, American Chemical Society

acid (NH_2–BDC). To load an optical imaging contrast agent, the MOF was treated with (1,3,5,7-tetramethyl- 4,4-difluoro-8-bromomethyl-4-bora-3a,4a-diaza-s-indacene (Br-BODIPY)) to yield a BODIPY-loaded particles with a loading capacity of 5.6–11.6 wt %. In addition, the ethoxysuccinato-cisplatin (ESCP), which is a prodrug of cisplatin, was also loaded into the MOF (Fig. 16). To improve the biological stability, the BODIPY- and ESCP-loaded MOF particles were also coated with silica shells using Na_2SiO_3 as the silica source to afford novel core–shell nanostructures. Laser scanning confocal microscopy images show that the BODIPY-loaded particles could cross the cell membrane and release the BODIPY dye inside of the cell, thus the fluorescence was present in cells incubated with BODIPY-loaded particles but absent in cells incubated without nanoparticles, suggesting that the MOF is an efficient platform for delivering an optical contrast agent. In addition, the modified BODIPY ligand was incubated with the cells, but no fluorescence was observed at all within the cells, indicating that the core–shell nanostructure was required for cell uptake. Treatment of HT-29 cells with

ESCP-loaded MOF particles shows appreciable cytotoxicity ($IC_{50} = 29\ \mu M$), which is slightly less cytotoxic than cisplatin under the same conditions ($IC_{50} = 20\ \mu M$). Further functionalization of silica-coated particles with silyl-derived c(RGDfK), which is a cyclic peptide known to target the $\alpha_v\beta_3$ integrin, shows that these particles has cytotoxicity ($IC_{50} = 21\ \mu M$) comparable to that of cisplatin. The generality of this approach could be utilized for the design of a wide range of nanomedical devices for theranostic applications.

5 Conclusions and Outlook

In the past decades, the growing research interest of chemists, physicists, and materials scientists has greatly accelerated the development of luminescent MOFs. Although luminescent MOFs are still in their infancy, the currently available results have unambiguously demonstrated that the design and construction of MOFs for luminescent functionality is very active, and hundreds of papers published in the last few years indicate that this interest is still increasing. The basic principles and the promises of luminescent MOFs as multifunctional materials have been established, as outlined in this chapter. On the one hand, comprehensive experimental studies and screening on luminescence properties on those reported and future-synthesized MOFs are still necessary to establish the database of luminescent MOFs. On the other hand, some in-depth studies on the mechanism of the luminescence behavior including the origin of the luminescence and structure-luminescent property relationship need to be carried out through the collaboration with theoretical scientists.

For construction of luminescent MOFs, it should be noted that the controllable synthesis of the MOFs is also a key issue to be addressed because of various factors affecting the formation process of the final products. The deliberate selection of organic linkers is very important to ensure the construction of MOF materials with luminescence. Rational control on structure, pore size, and functional sites within luminescent MOFs will be enforced to target highly sensitive and selective luminescent-sensing MOFs. Further efforts should be focused on the construction of porous luminescent MOFs with multifunctional sites to collaboratively induce their preferential and highly sensitive and selective binding with different ions. If the luminescent sensing could be combined with the sieving function of the micropores to allow small molecules to go through, while excluding larger species, it would be possible to obtain some highly sensitive sensors for small molecules.

In most of the luminescent MOFs, the sensing functionality is fulfilled based on the change of the luminescence intensity. The simple yet sophisticated ratiometric sensing approach is one of the breakthroughs on the exploration of functional luminescent MOF sensors. This approach makes the luminescent sensing independent of the concentration of the sample and of the drifts of the optoelectronic system including excitation source and detectors, thus overcoming the main drawbacks of the intensity-based measurements of only one transition.

Nanoscale luminescent MOFs will expect to be fabricated into thin films for their straightforward and instant sensing devices in the future. Additionally, nanoscale luminescent MOFs have bright future in the fields of cell imaging, as well as drug delivery monitoring and treatment, and will be extensively pursued for theranostic nanomedicine in the near future. Clearly, it would be interesting to utilize up-conversion luminescent MOFs in photodynamic therapy because the high loading capacity and controllable drug release of MOFs can facilitate the delivery and release of the photodynamic therapy drug. Moreover, the attractive multifunctionality of MOFs makes it possible to combine the up-conversion luminescence and drug delivery in a single component. In this regard, close multidisciplinary collaboration among chemists, materials scientists, biomedical scientists, and bioengineers will certainly facilitate the implementation of some promising luminescent MOF materials and technologies for practical applications for the theranostic nanomedicine.

Acknowledgment This work was supported by the National Natural Science Foundation of China (Nos. 51010002, 51272229, 51272231 and 51229201) and the Award AX-1730 from Welch Foundation (BC).

References

1. Zhou H-C, Long JR, Yaghi OM (2012) Introduction to metal-organic frameworks. Chem Rev 112:673–674
2. Suh MP, Park HJ, Prasad TK et al (2012) Hydrogen storage in metal-organic frameworks. Chem Rev 112:782–835
3. Sumida K, Rogow DL, Mason JA et al (2012) Carbon dioxide capture in metal-organic frameworks. Chem Rev 112:724–781
4. Li J-R, Sculley J, Zhou H-C (2012) Metal-organic frameworks for separations. Chem Rev 112:869–932
5. Yoon M, Srirambalaji R, Kim K (2012) Homochiral metal-organic frameworks for asymmetric heterogeneous catalysis. Chem Rev 112:1196–1231
6. Allendorf MD, Bauer CA, Bhakta RK et al (2009) Luminescent metal-organic frameworks. Chem Soc Rev 38:1330–1352
7. Cui Y, Yue Y, Qian G et al (2012) Luminescent functional metal-organic frameworks. Chem Rev 112:1126–1162
8. Chen BL, Xiang SC, Qian GD (2010) Metal-organic frameworks with functional pores for recognition of small molecules. Acc Chem Res 43:1115–1124
9. Kurmoo M (2009) Magnetic metal-organic frameworks. Chem Soc Rev 38:1353–1379
10. Horcajada P, Gref R, Baati T et al (2012) Metal-organic frameworks in biomedicine. Chem Rev 112:1232–1268
11. Murray LJ, Dinca M, Long JR (2009) Hydrogen storage in metal-organic frameworks. Chem Soc Rev 38:1294–1314
12. Férey G (2008) Hybrid porous solids: past, present, future. Chem Soc Rev 37:191–214
13. Czaja AU, Trukhan N, Müller U (2009) Industrial applications of metal-organic frameworks. Chem Soc Rev 38:1284
14. Zhang J-P, Zhang Y-B, Lin J-B et al (2012) Metal azolate frameworks: from crystal engineering to functional materials. Chem Rev 112:1001–1033

15. Stock N, Biswas S (2012) Synthesis of metal-organic frameworks (MOFs): routes to various mof topologies, morphologies, and composites. Chem Rev 112:933–969
16. Li H, Eddaoudi M, O'Keeffe M et al (1999) Design and synthesis of an exceptionally stable and highly porous metal-organic framework. Nature 402:276–279
17. Chen B, Eddaoudi M, Hyde ST (2001) Interwoven metal-organic framework on a periodic minimal surface with extra-large pores. Science 291:1021–1023
18. Rocha J, Carlos LD, Paz FAA et al (2011) Luminescent multifunctional lanthanides-based metal-organic frameworks. Chem Soc Rev 40:926
19. Kreno LE, Leong K, Farha OK et al (2012) Metal-organic framework materials as chemical sensors. Chem Rev 112:1105–1125
20. Jiang H-L, Xu Q (2011) Porous metal-organic frameworks as platforms for functional applications. Chem Commun 47:3351–3370
21. Eddaoudi M, Kim J, Rosi N et al (2002) Systematic design of pore size and functionality in isoreticular mofs and their application in methane storage. Science 295:469–472
22. Chen BL, Ma SQ, Zapata F et al (2007) Rationally designed micropores within a metal-organic framework for selective sorption of gas molecules. Inorg Chem 46:1233–1236
23. Chen BL, Liang CD, Yang J et al (2006) A microporous metal-organic framework for gas-chromatographic separation of alkanes. Angew Chem Int Ed 45:1390–1393
24. Wang RM, Zhang J, Li LJ (2010) A 2d metal-organic framework with a flexible cyclohexane-1,2,5,6-tetracarboxylic acid ligand: synthesis, characterization and photoluminescent property. J Mol Struct 970:14–18
25. Bai HY, Ma JF, Yang J et al (2010) Eight two-dimensional and three-dimensional metal organic frameworks based on a flexible tetrakis(imidazole) ligand: synthesis, topological structures, and photo luminescent properties. Cryst Growth Des 10:1946–1959
26. Ren P, Liu M-L, Zhang J et al (2008) 1d, 2d and 3d luminescent zinc(ii) coordination polymers assembled from varying flexible thioether ligands. Dalton Trans 4711–4713
27. Chai X-C, Sun Y-Q, Lei R et al (2010) A series of lanthanide frameworks with a flexible ligand, n, n′-diacetic acid imidazolium, in different coordination modes. Cryst Growth Des 10:658–668
28. Li D-P, Zhou X-H, Liang X-Q et al (2010) Novel structural diversity of triazolate-based coordination polymers generated solvothermally with anions. Cryst Growth Des 10: 2136–2145
29. Jing X, Meng H, Li G et al (2010) Construction of three metal-organic frameworks based on multifunctional T-shaped tripodal ligands, H_3pyimdc. Cryst Growth Des 10:3489–3495
30. Su Z, Chen S-S, Fan J et al (2010) Highly connected three-dimensional metal-organic frameworks based on polynuclear secondary building units. Cryst Growth Des 10:3675–3684
31. Ma LQ, Abney C, Lin WB (2009) Enantioselective catalysis with homochiral metal-organic frameworks. Chem Soc Rev 38:1248–1256
32. Li JR, Kuppler RJ, Zhou HC (2009) Selective gas adsorption and separation in metal-organic frameworks. Chem Soc Rev 38:1477–1504
33. Long JR, Yaghi OM (2009) The pervasive chemistry of metal-organic frameworks. Chem Soc Rev 38:1213–1214
34. Tian Y-Q, Chen Z-X, Weng L-H et al (2004) Two polymorphs of cobalt(II) imidazolate polymers synthesized solvothermally by using one organic template n, *n*-dimethylacetamide. Inorg Chem 43:4631–4635
35. Huang X-C, Lin Y-Y, Zhang J-P et al (2006) Ligand-directed strategy for zeolite-type metal-organic frameworks: Zinc(II) imidazolates with unusual zeolitic topologies. Angew Chem Int Ed 45:1557–1559
36. Fang Q, Zhu G, Xue M et al (2005) A metal-organic framework with the zeolite MTN topology containing large cages of volume 2.5 nm3. Angew Chem Int Ed 44:3845–3848
37. Tian Y-Q, Cai C-X, Ji Y et al (2002) $[Co_5(im)_{10}\cdot 2MB]\infty$: a metal-organic open-framework with zeolite-like topology Angew Chem Int Ed 41:1384–1386

38. Hayashi H, Côté AP, Furukawa H et al (2007) Zeolite a imidazolate frameworks. Nat Mater 6:501–506
39. Wang B, Côté AP, Furukawa H et al (2008) Colossal cages in zeolitic imidazolate frameworks as selective carbon dioxide reservoirs. Nature 453:207–211
40. MacGillivray L (2010) Metal-organic frameworks design and application. Wiley, New York
41. Klimakow M, Klobes P, Thünemann AF et al (2010) Mechanochemical synthesis of metal-organic frameworks: a fast and facile approach toward quantitative yields and high specific surface areas. Chem Mater 22:5216–5221
42. Yuan W, Friščić T, Apperley D et al (2010) High reactivity of metal-organic frameworks under grinding conditions: parallels with organic molecular materials. Angew Chem Int Ed 49:3916–3919
43. Braga D, Grepioni F, Maini L et al (2011) Solid-state reactivity of copper(I) iodide: luminescent 2D-coordination polymers of CuI with saturated bidentate nitrogen bases. New J Chem 35:339–344
44. Friščić T, Reid DG, Halasz I et al (2009) Ion- and liquid-assisted grinding: Improved mechanochemical synthesis of metal-organic frameworks reveals salt inclusion and anion templating. Angew Chem Int Ed 49:712–715
45. Schlesinger M, Schulze S, Hietschold M et al (2010) Evaluation of synthetic methods for microporous metal-organic frameworks exemplified by the competitive formation of $Cu_2(BTC)_3(H_2O)_3$ and $Cu_2(BTC)(OH)(H_2O)$. Microporous Mesoporous Mater 132:121–127
46. Yuan WB, Garay AL, Pichon A et al (2010) Study of the mechanochemical formation and resulting properties of an archetypal mof: $Cu_3(BTC)(_2)$ (btc = 1,3,5-benzenetricarboxylate). CrystEngComm 12:4063–4065
47. Pichon A, Lazuen-Garay A, James SL (2006) Solvent-free synthesis of a microporous metal-organic framework. CrystEngComm 8:211–214
48. Kole GK, Koh LL, Lee SY et al (2010) A new ligand for metal-organic framework and co-crystal synthesis: mechanochemical route to rctt-1,2,3,4-tetrakis-(4′-carboxyphenyl)-cyclobutane. Chem Commun 46:3660–3662
49. Fujii K, Garay AL, Hill J et al (2010) Direct structure elucidation by powder X-ray diffraction of a metal-organic framework material prepared by solvent-free grinding. Chem Commun 46:7572–7574
50. Son W-J, Kim J, Kim J, et al. (2008) Sonochemical synthesis of MOF-5. Chem Commun 44:6336–6338
51. Srivastava DN, Pol VG, Palchik O et al (2005) Preparation of stable porous nickel and cobalt oxides using simple inorganic precursor, instead of alkoxides, by a sonochemical technique. Ultrason Sonochem 12:205–212
52. Wang T, Lu XM, Han L et al (2008) Helical nanostructure of tubular metal-organic complex synthesized by sonochemical process. Sci China Ser B Chem 51:971–975
53. Jung DW, Yang DA, Kim J et al (2010) Facile synthesis of MOF-177 by a sonochemical method using 1-methyl-2-pyrrolidinone as a solvent. Dalton Trans 39:2883–2887
54. Sabouni R, Kazemian H, Rohani S (2010) A novel combined manufacturing technique for rapid production of IRMOF-1 using ultrasound and microwave energies. Chem Eng J 165: 966–973
55. Jung D-W, Yang D-A, Kim J et al (2010) Facile synthesis of MOF-177 by a sonochemical method using 1-methyl-2-pyrrolidinone as a solvent. Dalton Trans 39:2883
56. Prior TJ, Yotnoi B, Rujiwatra A (2011) Microwave synthesis and crystal structures of two cobalt-4,4′-bipyridine-sulfate frameworks constructed from 1-D coordination polymers linked by hydrogen bonding. Polyhedron 30:259–268
57. Khan NA, Haque E, Jhung SH (2010) Rapid syntheses of a metal-organic framework material $Cu_3(BTC)_2(H_2O)_3$ under microwave: a quantitative analysis of accelerated syntheses. PCCP 12:2625–2631

58. Haque E, Khan NA, Park JH et al (2010) Synthesis of a metal-organic framework material, iron terephthalate, by ultrasound, microwave, and conventional electric heating: a kinetic study. Chem Eur J 16:1046–1052
59. Lu CM, Liu J, Xiao KF et al (2010) Microwave enhanced synthesis of MOF-5 and its CO_2 capture ability at moderate temperatures across multiple capture and release cycles. Chem Eng J 156:465–470
60. Bux H, Liang FY, Li YS et al (2009) Zeolitic imidazolate framework membrane with molecular sieving properties by microwave-assisted solvothermal synthesis. J Am Chem Soc 131:16000
61. Liu HK, Tsao TH, Zhang YT et al (2009) Microwave synthesis and single-crystal-to-single-crystal transformation of magnesium coordination polymers exhibiting selective gas adsorption and luminescence properties. CrystEngComm 11:1462–1468
62. Yoo Y, Lai ZP, Jeong HK (2009) Fabrication of MOF-5 membranes using microwave-induced rapid seeding and solvothermal secondary growth. Microporous Mesoporous Mater 123: 100–106
63. Sonnauer A, Stock N (2008) High-throughput and microwave investigation of rare earth phosphonatoethanesulfonates-Ln(O_3P-C_2H_4-SO_3) (Ln = Ho, Er, Tm, Yb, Lu, Y). J Solid State Chem 181:3065–3070
64. Wang XF, Zhang YB, Huang H et al (2008) Microwave-assisted solvothermal synthesis of a dynamic porous metal-carboxylate framework. Crystal Growth Design 8:4559–4563
65. Ni Z, Masel RI (2006) Rapid production of metal-organic frameworks via microwave-assisted solvothermal synthesis. J Am Chem Soc 128:12394–12395
66. Centrone A, Harada T, Speakman S et al (2010) Facile synthesis of vanadium metal-organic frameworks and their magnetic properties. Small 6:1598–1602
67. Carné A, Carbonell C, Imaz I et al (2011) Nanoscale metal-organic materials. Chem Soc Rev 40:291–305
68. Jahan M, Bao QL, Yang JX et al (2010) Structure-directing role of graphene in the synthesis of metal-organic framework nanowire. J Am Chem Soc 132:14487–14495
69. Ma M, Zacher D, Zhang XN et al (2011) A method for the preparation of highly porous, nanosized crystals of isoreticular metal-organic frameworks. Crystal Growth Design 11: 185–189
70. Ostermann R, Cravillon J, Weidmann C et al (2011) Metal-organic framework nanofibers via electrospinning. Chem Commun 47:442–444
71. Li YS, Bux H, Feldhoff A et al (2010) Controllable synthesis of metal-organic frameworks: from mof nanorods to oriented mof membranes. Adv Mater 22:3322
72. Lin W, Rieter WJ, Taylor KML (2009) Modular synthesis of functional nanoscale coordination polymers. Angew Chem Int Ed 48:650–658
73. Rieter WJ, Taylor KML, An H et al (2006) Nanoscale metal-organic frameworks as potential multimodal contrast enhancing agents. J Am Chem Soc 128:9024–9025
74. Taylor KML, Rieter WJ, Lin W (2008) Manganese-based nanoscale metal-organic frameworks for magnetic resonance imaging. J Am Chem Soc 130:14358–14359
75. Xu H, Liu F, Cui Y et al (2011) A luminescent nanoscale metal-organic framework for sensing of nitroaromatic explosives. Chem Commun 47:3153–3155
76. Taylor KML, Jin A, Lin W (2008) Surfactant-assisted synthesis of nanoscale gadolinium metal-organic frameworks for potential multimodal imaging. Angew Chem Int Ed 47: 7722–7725
77. Zhou W, Wu H, Yildirim T (2008) Enhanced H_2 adsorption in isostructural metal-organic frameworks with open metal sites: strong dependence of the binding strength on metal ions. J Am Chem Soc 130:15268–15269
78. Xiang S, Zhou W, Gallegos JM et al (2009) Exceptionally high acetylene uptake in a microporous metal-organic framework with open metal sites. J Am Chem Soc 131: 12415–12419

79. Hasegawa S, Horike S, Matsuda R et al (2007) Three-dimensional porous coordination polymer functionalized with amide groups based on tridentate ligand: selective sorption and catalysis. J Am Chem Soc 129:2607–2614
80. Hwang YK, Hong D-Y, Chang J-S et al (2008) Amine grafting on coordinatively unsaturated metal centers of MOFs: consequences for catalysis and metal encapsulation. Angew Chem Int Ed 47:4144–4148
81. Xiang S, Zhou W, Zhang Z et al (2010) Open metal sites within isostructural metal-organic frameworks for differential recognition of acetylene and extraordinarily high acetylene storage capacity at room temperature. Angew Chem Int Ed 49:4615–4618
82. Cohen SM (2010) Modifying mofs: new chemistry, new materials. Chem Sci 1:32–36
83. Tanabe KK, Wang Z, Cohen SM (2008) Systematic functionalization of a metal-organic framework via a postsynthetic modification approach. J Am Chem Soc 130:8508–8517
84. Wang Z, Cohen SM (2007) Postsynthetic covalent modification of a neutral metal-organic framework. J Am Chem Soc 129:12368–12369
85. Nguyen JG, Cohen SM (2010) Moisture-resistant and superhydrophobic metal-organic frameworks obtained via postsynthetic modification. J Am Chem Soc 132:4560–4561
86. Wang Z, Tanabe KK, Cohen SM (2010) Tuning hydrogen sorption properties of metal-organic frameworks by postsynthetic covalent modification. Chem Eur J 16:212–217
87. Taylor-Pashow KML, Rocca JD, Xie Z et al (2009) Postsynthetic modifications of iron-carboxylate nanoscale metal-organic frameworks for imaging and drug delivery. J Am Chem Soc 131:14261–14263
88. Goto Y, Sato H, Shinkai S et al (2008) "Clickable" metal-organic framework. J Am Chem Soc 130:14354–14355
89. Savonnet M, Bazer-Bachi D, Bats N et al (2010) Generic postfunctionalization route from amino-derived metal-organic frameworks. J Am Chem Soc 132:4518–4519
90. Gadzikwa T, Farha OK, Malliakas CD et al (2009) Selective bifunctional modification of a non-catenated metal-organic framework material via "click" chemistry. J Am Chem Soc 131:13613–13615
91. Berezin MY, Achilefu S (2010) Fluorescence lifetime measurements and biological imaging. Chem Rev 110:2641–2684
92. Li X, Wang X-W, Zhang Y-H (2008) Blue photoluminescent 3d Zn(II) metal-organic framework constructing from pyridine-2,4,6-tricarboxylate. Inorg Chem Commun 11: 832–834
93. Fang Q, Zhu G, Xue M et al (2006) Structure, luminescence, and adsorption properties of two chiral microporous metal-organic frameworks. Inorg Chem 45:3582–3587
94. Shustova NB, McCarthy BD, Dinca M (2011) Turn-on fluorescence in tetraphenylethylene-based metal-organic frameworks: an alternative to aggregation-induced emission. J Am Chem Soc 133:20126–20129
95. Moore EG, Samuel APS, Raymond KN (2009) From antenna to assay: lessons learned in lanthanide luminescence. Acc Chem Res 42:542–552
96. Bünzli J-CG, Piguet C (2002) Lanthanide-containing molecular and supramolecular polymetallic functional assemblies. Chem Rev 102:1897–1928
97. Sabbatini N, Guardigli M, Lehn J-M (1993) Luminescent lanthanide complexes as photochemical supramolecular devices. Coord Chem Rev 123:201–228
98. Binnemans K (2009) Lanthanide-based luminescent hybrid materials. Chem Rev 109: 4283–4374
99. Eliseeva SV, Pleshkov DN, Lyssenko KA et al (2010) Highly luminescent and triboluminescent coordination polymers assembled from lanthanide β-diketonates and aromatic bidentate o-donor ligands. Inorg Chem 49:9300–9311
100. de Lill DT, de Bettencourt-Dias A, Cahill CL (2007) Exploring lanthanide luminescence in metal-organic frameworks: synthesis, structure, and guest-sensitized luminescence of a mixed europium/terbium-adipate framework and a terbium-adipate framework. Inorg Chem 46:3960–3965

101. Soares-Santos PCR, Cunha-Silva L, Paz FAA et al (2008) Photoluminescent 3d lanthanide-organic frameworks with 2,5-pyridinedicarboxylic and 1,4-phenylenediacetic acids. Cryst Growth Des 8:2505–2516
102. Li C, Lin J (2010) Rare earth fluoride nano-/microcrystals: synthesis, surface modification and application. J Mater Chem 20:6831
103. Carlos LD, Ferreira RAS, de Zea Bermudez V et al (2011) Progress on lanthanide-based organic–inorganic hybrid phosphors. Chem Soc Rev 40:536
104. Yang J, Yue Q, Li G-D et al (2006) Structures, photoluminescence, up-conversion, and magnetism of 2d and 3d rare-earth coordination polymers with multicarboxylate linkages. Inorg Chem 45:2857–2865
105. Weng D, Zheng X, Jin L (2006) Assembly and upconversion properties of lanthanide coordination polymers based on hexanuclear building blocks with (μ3-OH) bridges. Eur J Inorg Chem 2006:4184–4190
106. Bünzli J-CG (2010) Lanthanide luminescence for biomedical analyses and imaging. Chem Rev 110: 2729–2755
107. Mahata P, Ramya KV, Natarajan S (2008) Pillaring of $CdCl_2$-like layers in lanthanide metal-organic frameworks: synthesis, structure, and photophysical properties. Chem Eur J 14: 5839–5850
108. Xu J, Su W, Hong M (2011) A series of lanthanide secondary building units based metal-organic frameworks constructed by organic pyridine-2,6-dicarboxylate and inorganic sulfate. Cryst Growth Des 11:337–346
109. Chandler BD, Yu JO, Cramb DT et al (2007) Series of lanthanide-alkali metal-organic frameworks exhibiting luminescence and permanent microporosity. Chem Mater 19: 4467–4473
110. Shi FN, Cunha-Silva L, Mafra L et al (2008) Interconvertable modular framework and layered lanthanide(III)-etidronic acid coordination polymers. J Am Chem Soc 130:150–167
111. Zou J-P, Peng Q, Wen Z et al (2010) Two novel metal-organic frameworks (MOFs) with (3,6)-connected net topologies: syntheses, crystal structures, third-order nonlinear optical and luminescent properties. Cryst Growth Des 10:2613–2619
112. Wang G-H, Li Z-G, Jia H-Q et al (2009) Metal-organic frameworks based on the pyridine-2,3-dicarboxylate and a flexible bispyridyl ligand: syntheses, structures, and photoluminescence. CrystEngComm 11:292–297
113. Feng R, Jiang F-L, Chen L et al (2009) A luminescent homochiral 3d Cd(II) framework with a threefold interpenetrating uniform net 86. Chem Commun 43:5296
114. Wang XW, Chen J-Z, Liu J-H (2007) Photoluminescent Zn(II) metal−organic frameworks built from tetrazole ligand: 2d four-connected regular honeycomb (4363)-net. Cryst Growth Des 7:1227–1229
115. An J, Shade CM, Chengelis-Czegan DA et al (2011) Zinc-adeninate metal-organic framework for aqueous encapsulation and sensitization of near-infrared and visible emitting lanthanide cations. J Am Chem Soc 133:1220–1223
116. Luo F, Batten SR (2010) Metal-organic framework (mof): Lanthanide(III)-doped approach for luminescence modulation and luminescent sensing. Dalton Trans 39:4485
117. Park YK, Choi SB, Kim H et al (2007) Crystal structure and guest uptake of a mesoporous metal-organic framework containing cages of 3.9 and 4.7 nm in diameter. Angew Chem Int Ed 46:8230–8233
118. Buso D, Jasieniak J, Lay MDH et al (2012) Highly luminescent metal-organic frameworks through quantum dot doping. Small 8:80–88
119. Fang Q-R, Zhu G-S, Jin Z et al (2007) Mesoporous metal-organic framework with rare etb topology for hydrogen storage and dye assembly. Angew Chem Int Ed 46:6638–6642
120. Rodríguez-Diéguez A, Salinas-Castillo A, Sironi A et al (2010) A chiral diamondoid 3D lanthanum metal-organic framework displaying blue-greenish long lifetime photoluminescence emission. CrystEngComm 12:1876

121. Rao X, Huang Q, Yang X et al (2012) Color tunable and white light emitting Tb^{3+} and Eu^{3+} doped lanthanide metal-organic framework materials. J Mater Chem 22:3210–3214
122. Feng X, Wang L-Y, Zhao J-S et al (2010) Series of anion-directed lanthanide-rigid-flexible frameworks: syntheses, structures, luminescence and magnetic properties. CrystEngComm 12:774
123. Zhang L-Z, Gu W, Li B et al (2007) $\{[Nd_4(ox)_4(NO_3)_2(OH)_2(H_2O)_2]\cdot 5H_2O\}_n$: a porous 3d lanthanide-based coordination polymer with a special luminescent property. Inorg Chem 46:622–624
124. Gl Z, Maury O, Thuéry P et al (2008) Structural diversity in neodymium bipyrimidine compounds with near infrared luminescence: from mono- and binuclear complexes to metal-organic frameworks. Inorg Chem 47:10398–10406
125. Zhu X, Lü J, Li X et al (2008) Syntheses, structures, near-infrared, and visible luminescence of lanthanide-organic frameworks with flexible macrocyclic polyamine ligands. Cryst Growth Des 8:1897–1901
126. Wang H-S, Zhao B, Zhai B et al (2007) Syntheses, structures, and photoluminescence of one-dimensional lanthanide coordination polymers with 2,4,6-pyridinetricarboxylic acid. Cryst Growth Des 7:1851–1857
127. Liao J-H, Tsai C-S, Lin T-K (2010) Syntheses, structural characterization and luminescent properties of $M_2(ATPA)_3(DMF)_2(H_2O)_2$ (M = Nd, Sm, Eu, Gd, Tb, Dy; ATPA = 2-aminoterephthalate, DMF = N, N-dimethylformamide). Inorg Chem Commun 13:286–289
128. Daiguebonne C, Kerbellec N, Guillou O et al (2008) Structural and luminescent properties of micro- and nanosized particles of lanthanide terephthalate coordination polymers. Inorg Chem 47:3700–3708
129. Han Y, Li X, Li L et al (2010) Structures and properties of porous coordination polymers based on lanthanide carboxylate building units. Inorg Chem 49:10781–10787
130. de Lill DT, Gunning NS, Cahill CL (2005) Toward templated metal−organic frameworks: synthesis, structures, thermal properties, and luminescence of three novel lanthanide−adipate frameworks. Inorg Chem 44:258–266
131. Chen B, Wang L, Xiao Y et al (2009) A luminescent metal-organic framework with Lewis basic pyridyl sites for the sensing of metal ions. Angew Chem Int Ed 48:500–503
132. Xiao Y, Cui Y, Zheng Q et al (2010) A microporous luminescent metal-organic framework for highly selective and sensitive sensing of Cu^{2+} in aqueous solution. Chem Commun 46: 5503
133. Chen B, Yang Y, Zapata F et al (2007) Luminescent open metal sites within a metal-organic framework for sensing small molecules. Adv Mater 19:1693–1696
134. Ma D, Wang W, Li Y et al (2010) In situ 2,5-pyrazinedicarboxylate and oxalate ligands synthesis leading to a microporous europium–organic framework capable of selective sensing of small molecules. CrystEngComm 12:4372
135. Lin Z-J, Xu B, Liu T-F et al (2010) A series of lanthanide metal-organic frameworks based on biphenyl-3,4′,5-tricarboxylate: syntheses, structures, luminescence and magnetic properties. Eur J Inorg Chem 2010:3842–3849
136. Zhu W-H, Wang Z-M, Gao S (2007) Two 3d porous lanthanide−fumarate−oxalate frameworks exhibiting framework dynamics and luminescent change upon reversible de- and rehydration. Inorg Chem 46:1337–1342
137. Harbuzaru BV, Corma A, Rey F et al (2008) Metal-organic nanoporous structures with anisotropic photoluminescence and magnetic properties and their use as sensors. Angew Chem Int Ed 47:1080–1083
138. Harbuzaru BV, Corma A, Rey F et al (2009) A miniaturized linear ph sensor based on a highly photoluminescent self-assembled europium(III) metal-organic framework. Angew Chem Int Ed 48:6476–6479

139. Yang X-P, Jones RA, Rivers JH et al (2007) Syntheses, structures and luminescent properties of new lanthanide-based coordination polymers based on 1,4-benzenedicarboxylate (bdc). Dalton Trans 35:3936
140. Sun Y-G, Yu W, Wang L et al (2010) Synthesis, structures, and luminescence of lanthanide coordination polymers constructed from benzimidazole-5,6-dicarboxylate and oxalate ligands. Inorg Chem Commun 13:479–483
141. Lu W-G, Jiang L, Feng X-L et al (2009) Three-dimensional lanthanide anionic metal−organic frameworks with tunable luminescent properties induced by cation exchange. Inorg Chem 48:6997–6999
142. Wong KL, Law GL, Yang YY et al (2006) A highly porous luminescent terbium–organic framework for reversible anion sensing. Adv Mater 18:1051–1054
143. Chen B, Wang L, Zapata F et al (2008) A luminescent microporous metal−organic framework for the recognition and sensing of anions. J Am Chem Soc 130:6718–6719
144. Xu H, Xiao Y, Rao X et al (2011) A metal-organic framework for selectively sensing of PO_4^{3-} anion in aqueous solution. J Alloys Compd 509:2552–2554
145. Lin Y-W, Jian B-R, Huang S-C et al (2010) Synthesis and characterization of three ytterbium coordination polymers featuring various cationic species and a luminescence study of a terbium analogue with open channels. Inorg Chem 49:2316–2324
146. Yang X, Rivers JH, McCarty WJ et al (2008) Synthesis and structures of luminescent ladder-like lanthanide coordination polymers of 4-hydroxybenzenesulfonate. New J Chem 32:790
147. Guo X, Zhu G, Sun F et al (2006) Synthesis, structure, and luminescent properties of microporous lanthanide metal−organic frameworks with inorganic rod-shaped building units. Inorg Chem 45:2581–2587
148. Chen B, Yang Y, Zapata F et al (2006) Enhanced near-infrared-luminescence in an erbium tetrafluoroterephthalate framework. Inorg Chem 45:8882–8886
149. Guo Z, Xu H, Su S et al (2011) A robust near infrared luminescent ytterbium metal-organic framework for sensing of small molecules. Chem Commun 47:5551–5553
150. White KA, Chengelis DA, Zeller M et al (2009) Near-infrared emitting ytterbium metal-organic frameworks with tunable excitation properties. Chem Commun 45:4506
151. Huang W, Wu D, Zhou P et al (2009) Luminescent and magnetic properties of lanthanide-thiophene-2,5-dicarboxylate hybrid materials. Cryst Growth Des 9:1361–1369
152. Lee EY, Jang SY, Suh MP (2005) Multifunctionality and crystal dynamics of a highly stable, porous metal−organic framework $[Zn_4O(NTB)_2]$. J Am Chem Soc 127:6374–6381
153. Feng PL, Perry Iv JJ, Nikodemski S et al (2010) Assessing the purity of metal−organic frameworks using photoluminescence: MOF-5, ZnO quantum dots, and framework decomposition. J Am Chem Soc 132:15487–15489
154. Zhang Z, Xiang S, Rao X et al (2010) A rod packing microporous metal-organic framework with open metal sites for selective guest sorption and sensing of nitrobenzene. Chem Commun 46:7205
155. Zhang Z, Xiang S, Zheng Q et al (2010) A rare uninodal 9-connected metal-organic framework with permanent porosity. Cryst Growth Des 10:2372–2375
156. Hou L, Lin Y-Y, Chen X-M (2008) Porous metal−organic framework based on μ4-oxo tetrazinc clusters: sorption and guest-dependent luminescent properties. Inorg Chem 47: 1346–1351
157. Lan A, Li K, Wu H et al (2009) A luminescent microporous metal-organic framework for the fast and reversible detection of high explosives. Angew Chem Int Ed 121:2370–2374
158. Zhang C, Che Y, Zhang Z et al (2011) Fluorescent nanoscale zinc(II)-carboxylate coordination polymers for explosive sensing. Chem Commun 47:2336–2338
159. Bauer CA, Timofeeva TV, Settersten TB et al (2007) Influence of connectivity and porosity on ligand-based luminescence in zinc metal-organic frameworks. J Am Chem Soc 129: 7136–7144
160. Jiang H-L, Liu B, Xu Q (2010) Rational assembly of d10 metal-organic frameworks with helical nanochannels based on flexible V-shaped ligand. Cryst Growth Des 10:806–811

161. Yang E-C, Zhao H-K, Ding B et al (2007) Four novel three-dimensional triazole-based zinc (II) metal-organic frameworks controlled by the spacers of dicarboxylate ligands: Hydrothermal synthesis, crystal structure, and luminescence properties. Cryst Growth Des 7:2009–2015
162. Guo H-D, Guo X-M, Batten SR et al (2009) Hydrothermal synthesis, structures, and luminescent properties of seven d10 metal-organic frameworks based on 9,9-dipropylfluorene-2,7-dicarboxylic acid (H_2DFDA). Cryst Growth Des 9:1394–1401
163. Chen F, Wu M-F, Liu G-N et al (2010) Zinc(II) and cadmium(II) coordination polymers based on 3-(5H-tetrazolyl)benzoate ligand with different coordination modes: hydrothermal syntheses, crystal structures and ligand-centered luminescence. Eur J Inorg Chem 2010: 4982–4991
164. Chen X-L, Gou L, Hu H-M et al (2008) New examples of metal coordination architectures of 4,4′-sulfonyldibenzoic acid: syntheses, crystal structure and luminescence. Eur J Inorg Chem 2008:239–250
165. Xue M, Zhu G, Zhang Y et al (2008) Rational design and control of the dimensions of channels in a series of 3d pillared metal-organic frameworks: synthesis, structures, adsorption, and luminescence properties. Cryst Growth Des 8:427–434
166. Jiang H-L, Tatsu Y, Lu Z-H et al (2010) Non-, micro-, and mesoporous metal-organic framework isomers: reversible transformation, fluorescence sensing, and large molecule separation. J Am Chem Soc 132:5586–5587
167. Qiu Y, Li Y, Peng G et al (2010) Cadmium metal-directed three-dimensional coordination polymers: in situ tetrazole ligand synthesis, structures, and luminescent properties. Cryst Growth Des 10:1332–1340
168. Liu H-J, Tao X-T, Yang J-X et al (2008) Three-dimensional metal-organic network architecture with large π-conjugated indolocarbazole derivative: synthesis, supramolecular structure, and highly enhanced fluorescence. Cryst Growth Des 8:259–264
169. Xue M, Zhu G, Li Y et al (2008) Structure, hydrogen storage, and luminescence properties of three 3d metal-organic frameworks with NbO and PtS topologies. Cryst Growth Des 8: 2478–2483
170. Wang M-S, Guo S-P, Li Y et al (2009) A direct white-light-emitting metal-organic framework with tunable yellow-to-white photoluminescence by variation of excitation light. J Am Chem Soc 131:13572–13573
171. Song L, Du S-W, Lin J-D et al (2007) A 3d metal-organic framework with rare 3-fold interpenetrating dia-g nets based on silver(I) and novel tetradentate imidazolate ligand: synthesis, structure, and possible ferroelectric property. Cryst Growth Des 7:2268–2271
172. Wu H-C, Thanasekaran P, Tsai C-H et al (2006) Self-assembly, reorganization, and photophysical properties of silver(I)-schiff-base molecular rectangle and polymeric array species. Inorg Chem 45:295–303
173. Bai Y, Gao H, Dang D-B et al (2010) A series of metal-organic frameworks based on polydentate schiff-base ligands derived from benzil dihydrazone: synthesis, crystal structures and luminescent properties. CrystEngComm 12:1422
174. Zhang T, Ji C, Wang K et al (2010) First halogen anion-bridged (MMX)n-type one-dimensional coordination polymer built upon d10−d10 dimers. Inorg Chem 49: 11069–11076
175. Jiang H, Ma J-F, Zhang W-L et al (2008) Metal-organic frameworks containing flexible bis (benzimidazole) ligands. Eur J Inorg Chem 2008:745–755
176. Bai Y, He G-J, Zhao Y-G et al (2006) Porous material for absorption and luminescent detection of aromatic molecules in water. Chem Commun 42:1530
177. Li M-X, Wang H, Liang S-W et al (2009) Solvothermal synthesis and diverse coordinate structures of a series of luminescent copper(I) thiocyanate coordination polymers based on n-heterocyclic ligands. Cryst Growth Des 9:4626–4633
178. Liu X, Huang K-L (2009) A 12-connected dodecanuclear copper cluster with yellow luminescence. Inorg Chem 48:8653–8655

179. Xia J, Zhang Z-J, Shi W et al (2010) Copper(I) cyanide coordination polymers constructed from bis(pyrazole-1-yl)alkane ligands: observation of the odd-even dependence in the structures. Cryst Growth Des 10:2323–2330
180. He J, Yin Y-G, Wu T et al (2006) Design and solvothermal synthesis of luminescent copper (I)-pyrazolate coordination oligomer and polymer frameworks. Chem Commun 42:2845
181. Xiong S, Wang S, Tang X et al (2011) Four new metal-organic frameworks constructed from H_2DBTDC-O_2 (H_2DBTDC-O_2=dibenzothiophene-5,5′-dioxide-3,7-dicarboxylic acid) ligand with guest-responsive photoluminescence. CrystEngComm 13:1646–1653
182. Wei Y, Yu Y, Wu K (2008) Highly stable five-coordinated Mn(II) polymer [Mn(hbidc)]n (hbidc = 1H-benzimidazole-5,6-dicarboxylate): crystal structure, antiferromegnetic property, and strong long-lived luminescence. Cryst Growth Des 8:2087–2089
183. Bo Q-B, Sun G-X, Geng D-L (2010) Novel three-dimensional pillared-layer Ln(III)−Cu (I) coordination polymers featuring spindle-shaped heterometallic building units. Inorg Chem 49:561–571
184. Sun Y-Q, Zhang J, Yang G-Y (2006) A series of luminescent lanthanide–cadmium–organic frameworks with helical channels and tubes. Chem Commun 45:4700
185. Zhao X-Q, Zhao B, Shi W et al (2009) Structures and luminescent properties of a series of ln–ag heterometallic coordination polymers. CrystEngComm 11:1261
186. Liu W, Li Z, Wang N et al (2011) A new family of 3d heterometallic 3d–4f organodisulfonate complexes based on the linkages of 2D $[ln(nds)(H_2O)]^+$ layers and $[Cu(ina)2]^-$ chains. CrystEngComm 13:138
187. Chandler BD, Cramb DT, Shimizu GKH (2006) Microporous metal−organic frameworks formed in a stepwise manner from luminescent building blocks. J Am Chem Soc 128: 10403–10412
188. Zhao B, Chen X-Y, Cheng P et al (2004) Coordination polymers containing 1d channels as selective luminescent probes. J Am Chem Soc 126:15394–15395
189. Zhao B, Chen X-Y, Chen Z et al (2009) A porous 3d heterometal-organic framework containing both lanthanide and high-spin Fe(II) ions. Chem Commun 45:3113
190. Gu X, Xue D (2007) Self-assembly of 3-d 4d-4f coordination frameworks based on 1-d inorganic heterometallic chains and linear organic linkers. CrystEngComm 9:471
191. Peng G, Qiu Y-C, Liu Z-H et al (2008) Construction of three isostructural 3d–4f microporous coordination frameworks based on mixed nicotinate and oxalate ligands. Inorg Chem Commun 11:1409–1411
192. Wang P, Ma J-P, Dong Y-B et al (2007) Tunable luminescent lanthanide coordination polymers based on reversible solid-state ion-exchange monitored by ion-dependent photo-induced emission spectra. J Am Chem Soc 129:10620–10621
193. Jin J, Niu S, Han Q et al (2010) Synthesis and structure of a series of new luminescent Ag–Ln coordination polymers and the influence of the introduction of an Ag(I) ion on nir luminescence from the Ln(III) centre. New J Chem 34:1176
194. Zhao B, Zhao XQ, Chen Z et al (2008) Structures and near-infrared luminescence of unique 4d–4f heterometal-organic frameworks (hmof). CrystEngComm 10:1144
195. Prasad TK, Rajasekharan MV (2009) Cerium(IV)-lanthanide(III)-pyridine-2,6-dicarboxylic acid system: coordination salts, chains, and rings. Inorg Chem 48:11543–11550
196. Guo H, Zhu Y, Qiu S et al (2010) Coordination modulation induced synthesis of nanoscale eu1-xtbx-metal-organic frameworks for luminescent thin films. Adv Mater 22:4190–4192
197. Liu K, You H, Zheng Y et al (2010) Facile and rapid fabrication of metal-organic framework nanobelts and color-tunable photoluminescence properties. J Mater Chem 20:3272
198. Cui Y, Xu H, Yue Y et al (2012) A luminescent mixed-lanthanide metal-organic framework thermometer. J Am Chem Soc 134:3979–3982
199. White KA, Chengelis DA, Gogick KA et al (2009) Near-infrared luminescent lanthanide mof barcodes. J Am Chem Soc 131:18069–18071
200. Xie Z, Ma L, deKrafft KE et al (2010) Porous phosphorescent coordination polymers for oxygen sensing. J Am Chem Soc 132:922–923

201. Stylianou KC, Heck R, Chong SY et al (2010) A guest-responsive fluorescent 3d microporous metal–organic framework derived from a long-lifetime pyrene core. J Am Chem Soc 132: 4119–4130
202. Sava DF, Rohwer LES, Rodriguez MA et al (2012) Intrinsic broad-band white-light emission by a tuned, corrugated metal-organic framework. J Am Chem Soc 134:3983–3986
203. Wibowo AC, Vaughn SA, Smith MD et al (2010) Novel bismuth and lead coordination polymers synthesized with pyridine-2,5-dicarboxylates: two single component "white" light emitting phosphors. Inorg Chem 49:11001–11008
204. Yang E-C, Li J, Ding B et al (2008) An eight-connected 3d lead(ii) metal-organic framework with octanuclear lead(ii) as a secondary building unit: synthesis, characterization and luminescent property. CrystEngComm 10:158–161
205. Hu R, Cai H, Luo J (2011) Synthesis, structure and luminescent property of a pillared-layer coordination polymer [pb(bpdc)] (bpdc = 4,4′-biphenyldicarboxylate). Inorg Chem Commun 14:433–436
206. D'Andrade BW, Forrest SR (2004) White organic light-emitting devices for solid-state lighting. Adv Mater 16:1585
207. Eliseeva SV, Bünzli J-CG (2010) Lanthanide luminescence for functional materials and bio-sciences. Chem Soc Rev 39:189
208. Dang S, Zhang J-H, Sun Z-M (2012) Tunable emission based on lanthanide(III) metal-organic frameworks: an alternative approach to white light. J Mater Chem 22:8868
209. Kozlowski H, Janicka-Klos A, Brasun J et al (2009) Copper, iron, and zinc ions homeostasis and their role in neurodegenerative disorders (metal uptake, transport, distribution and regulation). Coord Chem Rev 253:2665–2685
210. Que EL, Domaille DW, Chang CJ (2008) Metals in neurobiology: probing their chemistry and biology with molecular imaging. Chem Rev 108:1517–1549
211. Qiu Y, Deng H, Mou J et al (2009) In situ tetrazole ligand synthesis leading to a microporous cadmium–organic framework for selective ion sensing. Chem Commun 45:5415
212. Xu H, Rao X, Gao J et al (2012) A luminescent nanoscale metal-organic framework with controllable morphologies for spore detection. Chem Commun 48:7377–7379
213. Takashima Y, Martínez VM, Furukawa S et al (2011) Molecular decoding using luminescence from an entangled porous framework. Nat Comms 2:168
214. Katz MJ, Ramnial T, Yu H-Z et al (2008) Polymorphism of $Zn[Au(CN)_2]_2$and its luminescent sensory response to NH_3 vapor. J Am Chem Soc 130:10662–10673
215. Habibagahi A, Mébarki Y, Sultan Y et al (2009) Water-based oxygen-sensor films. ACS Appl Mater Interfaces 1:1785–1792
216. Li Y, Zhang S, Song D (2013) A luminescent metal-organic framework as a turn-on sensor for DMF vapor. Angew Chem Int Ed 52:710–713
217. Miyata K, Konno Y, Nakanishi T et al (2013) Chameleon luminophore for sensing temperatures: control of metal-to-metal and energy back transfer in lanthanide coordination polymers. Angew Chem Int Ed 52:6413–6416
218. Cadiau A, Brites CD, Costa PM et al (2013) Ratiometric nanothermometer based on an emissive Ln-organic framework. ACS Nano 7:7213–7218.

Struct Bond (2014) 157: 89–104
DOI: 10.1007/430_2013_131

Published online: 7 September 2013

Metal-Organic Frameworks for Photocatalysis

Teng Zhang and Wenbin Lin

Abstract Metal-organic frameworks (MOFs) have recently been investigated as a platform for developing single-site photocatalysts and multifunctional photocatalytic assemblies. This chapter aims to summarize the recent progress on photocatalysis with MOFs, including hydrogen evolution, carbon dioxide reduction, and degradation and transformation of organic compounds.

Keywords CO_2 reduction · Metal-organic frameworks · Photocatalysis · Water splitting

Contents

T. Zhang and W. Lin (✉)
Department of Chemistry, The University of Chicago, 929 E. 57th Street, Chicago, IL 60637, USA
e-mail: wenbinlin@uchicago.edu

1 Introduction

With the dwindling fossil fuel reserve and increasing global energy demand, solar energy is the best alternative energy source. Significant efforts have been devoted to harvesting and storing energy from the sunlight. Photochemical processes with net energy uphill reactions ($\Delta G > 0$), such as water splitting and carbon dioxide reduction, have been proposed to transform and store sunlight energy in the form of chemical bonds. Green plants have developed effective but complex systems to practice these transformations, known as photosynthesis, to convert carbon dioxide and water into carbohydrates. Inspired by natural photosynthesis, scientists have strived to develop inorganic materials and devices for these photochemical transformations. Since the first report of UV light-driven water splitting in 1972 [1], different materials and devices have been demonstrated [2–5], yet it still remains a great challenge to develop a stable, effective, and affordable system that can utilize the full solar spectrum for solar energy conversion.

Chemists have also tried to mimic the molecular photochemical systems in green plants. Various molecular dyes as well as catalysts have been developed [6–11]. There are however no efficient strategies to combine these functional components effectively into hierarchical assemblies for artificial photosynthesis. Metal-organic frameworks (MOFs) provide a potential solution to this challenge. Constructed from well-defined metal/metal cluster nodes and molecular building blocks, MOFs have shown potential applications in many different areas [12–19] and offer a possibility to incorporate different functional components into a single solid. In particular, MOFs allow hierarchical assembly of multiple functionalities into a single solid material to potentially achieve artificial photosynthesis.

Light also serves as a means to activate organic molecules to undergo transformations that cannot be carried out under typical thermally activated reaction conditions. Although photocatalysis has long been proposed as an important methodology, the application is limited because most organic substrates only absorb UV photons, leaving the majority of the solar spectrum unused. Molecular dyes can be used to absorb lower-energy photons to activate organic substrates, leading to many novel photocatalytic reactions driven with visible light [20–24]. Photoactive MOFs can also serve as single-site photocatalysts and take advantage of the solid nature of MOF catalysts to facilitate recovery and reuse of expensive photocatalysts thereby reducing contamination by the photocatalysts which usually contain heavy metals. In this chapter, we will summarize recent progress on photocatalysis with MOFs, including water splitting, carbon dioxide reduction, and organic transformations.

2 Hydrogen Evolution with MOFs

During the past few decades, semiconducting materials and molecular complexes have been developed as photosensitizers or photocatalysts for splitting water into hydrogen and oxygen. Several MOFs have also been reported to serve as

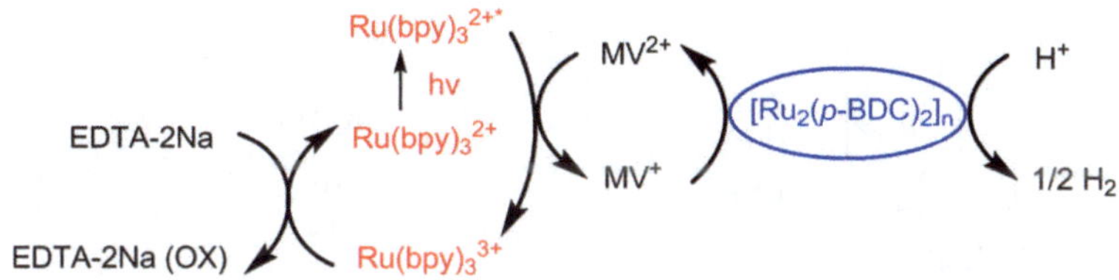

Fig. 1 Proposed photocatalytic cycle of hydrogen evolution from water using the $[Ru_2(p\text{-BDC})_2]_n$, $Ru(bpy)_3^{2+}$, MV^{2+}, and Na_2EDTA combination under visible light irradiation. Reprinted with permission from [25]. Copyright 2009 Royal Society of Chemistry

photosensitizers or catalysts for sacrificial proton reduction reactions in the past few years. In 2009, Mori and coworkers reported a porous MOF $[Ru_2(p\text{-BDC})_2]_n$ (**1**, *p*-BDC = *p*-benzenedicarboxylate) that catalyzes visible light-driven proton reduction with the $Ru(bpy)_3^{2+}/MV^{2+}$/EDTA system [25]. **1** adopts in a 2-D square grid structure constructed from $Ru_2(COO)_4$ paddle-wheel building blocks and linear *p*-BDC linkers, with 1-D channels running perpendicular to the 2-D sheets. In the system, **1** serves as the catalyst, $Ru(bpy)_3^{2+}$ acts as a photosensitizer, MV^{2+} acts as an electron relay, and Na_2EDTA acts as a sacrificial reductant. The catalytic proton reduction reaction gives a turnover number (TON) of 8.16 per Ru atom in a 4 h period under irradiation. The proposed catalytic cycle starts with the electron transfer between the excited state of $Ru(bpy)_3^{2+}$ and the electron relay MV^{2+} to yield $Ru(bpy)_3^{3+}$ and MV^+, while the former is reduced back to $Ru(bpy)_3^{2+}$ by Na_2EDTA and the latter transfers an electron to the catalyst **1** to reduce proton to hydrogen (Fig. 1). Although the proposed mechanism seems plausible, the authors provided little experimental evidence to support this proposal.

The Zr-based MOFs UiO-66 and UiO-66(NH_2), first developed by Lillerud and coworkers [26], have also been reported for photocatalytic water reduction. In 2010, García and coworkers reported hydrogen evolution catalyzed by the MOFs with a UV irradiation (>300 nm) and methanol as the sacrificial electron donor [27]. Platinum nanoparticles were also added as a cocatalyst. The presence of amino group shifts the absorption band to >300 nm and makes the photocatalysis possible. However, the need for UV photons presents a significant drawback for this experimental design.

In early 2012, Lin and coworkers reported a MOF assembly containing molecular phosphor building blocks and platinum nanoparticles for hydrogen evolution [28]. Iridium-based molecular photosensitizers bis(4-phenyl-2-pyridine)(5,5′-dicarboxylate)-2,2′-bipyridine)-iridium(III) chloride (**L_1**) or bis(4-phenyl-2-pyridine)(5,5′-di(4-phenylcarboxylate)-2,2′-bipyridine)-iridium(III) chloride (**L_2**) were used as building blocks, in combination with $Zr_6(O)_4(OH)_4$ SBUs, to construct water-stable MOFs **2** or **3** (Fig. 2). Platinum nanoparticles were deposited in the MOF channels via in situ MOF-mediated photoreduction of platinum precursor (Fig. 3). Higher nanoparticle loading is found in **3** than **2** as **3** has a larger void space and a higher phosphor concentration. Both Pt@MOF assemblies exhibited high catalytic activity for proton reduction with triethylamine as the sacrificial reductant. Pt@**2** and Pt@**3** exhibited a

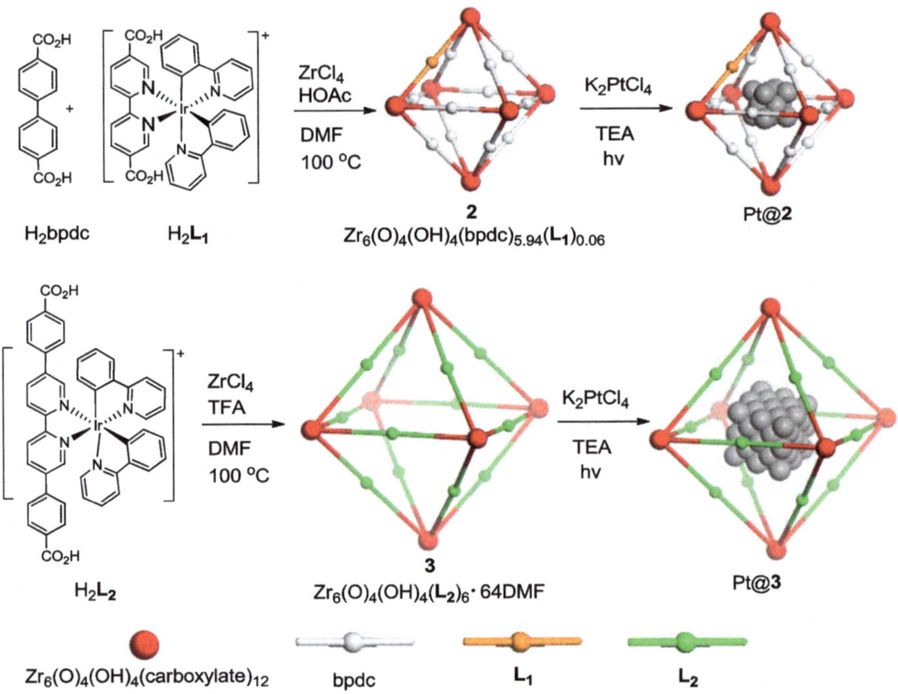

Fig. 2 Synthesis of phosphorescent Zr-carboxylate MOFs and subsequent loading of Pt NPs inside MOFs **2** and **3**. Reprinted with permission from [28]. Copyright 2012 American Chemical Society

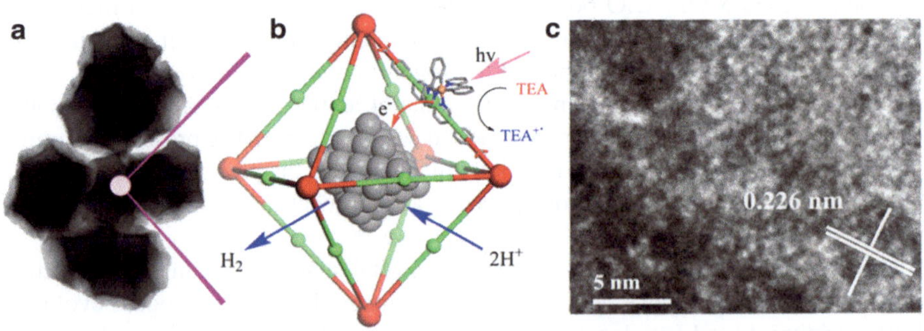

Fig. 3 (**a**) TEM images of **Pt@3**. (**b**) The synergistic photocatalytic hydrogen generation process via photoinjection of electron from the light-harvesting MOF on to the Pt NPs. (**c**) HRTEM images of a powdery sample of **Pt@4** that shows the lattice fringes of the Pt particles, with d-spacing matching that of the Pt{111} plane. Reprinted with permission from [28]. Copyright 2012 American Chemical Society

TON of 3,400 and 7,000 per Ir for hydrogen evolution over an illumination time of 48 h, respectively. These TONs are almost five times as high as the homogeneous control. It is believed that facile electron transfer between the photoreduced Ir-complexes and the Pt nanoparticles enhances the catalytic performance of the Pt@MOFs over their homogeneous counterparts. The catalysts can be recycled and reused for at least three times though significant leaching of the iridium complexes was also observed after the 48 h reactions.

Porphyrin-based ligands were also used to construct photoactive MOFs due to their interesting photophysical properties. In 2012, Rosseinsky and coworkers reported a water-stable porphyrin MOF (Al-PMOF) and its catalytic activity for proton reduction [29]. They first tested the Al-PMOF/MV^{2+}/EDTA/Pt system, which is analogous to the one used by Mori et al. [25] but using Al-PMOF instead of $Ru(bpy)_3^{2+}$ as the photosensitizer. However, very low activity was observed. When MV^{2+} was removed from the system, the hydrogen generation activity increased by almost one order of magnitude and an approximate TON of 0.7 H_2/porphyrin was achieved over a period of 6 h. The authors believed that the slow diffusion of methyl viologen through the MOF channels results in ineffective electron transfer and leads to low activity. Stability of the Al-PMOF was supported by powder XRD after reaction and further confirmed by the absence of porphyrin leaching into the solution after reaction.

3 Carbon Dioxide Reduction with MOFs

Photochemical reduction of CO_2 is another important strategy for solar energy conversion. Not only does it harvest energy from sunlight to produce valuable organic compounds but also helps to reduce the CO_2 level in the atmosphere. In natural photosynthesis, CO_2 is reduced by photochemically generated NADPH in photosystem I [30]. Scientists have developed a number of molecular compounds, semiconductors, and metal-incorporated zeolites for artificial photochemical CO_2 reduction [31–34].

Lin and coworkers reported in 2011 the synthesis of MOF **4** by doping [$Re^I(CO)_3$(5,5′-dcbpy)Cl] (**L_3**), a derivative of the CO_2 reduction catalyst $Re^I(CO)_3$(bpy)X [31, 36, 37], into the UiO-67 framework at a loading of 4 wt% [35]. The synthesis of **4** took advantage of the matching ligand lengths of [$Re^I(CO)_3$(5,5′-dcbpy)Cl] and bpdc, while **4** showed the same PXRD pattern as UiO-67 (Scheme 1 and Fig. 4c). Photocatalytic CO_2 reduction was carried out in a CO_2-saturated acetonitrile solution with triethylamine as the sacrificial reductant and **4** as the catalyst. **4** selectively reduced CO_2 to CO with a TON of 10.9 over a period of 12 h, almost three times higher than the homogeneous complex (Fig. 4a). The higher activity was believed to be a result of blocking the bimolecular decomposition due to site isolation of the [$Re^I(CO)_3$(5,5′-dcbpy)Cl] molecules. Moreover, **4** exhibited high selectivity towards CO. The molar amount of CO was ten times as much as H_2 generated in the first 6 h, and no formic acid or methanol was detected during the

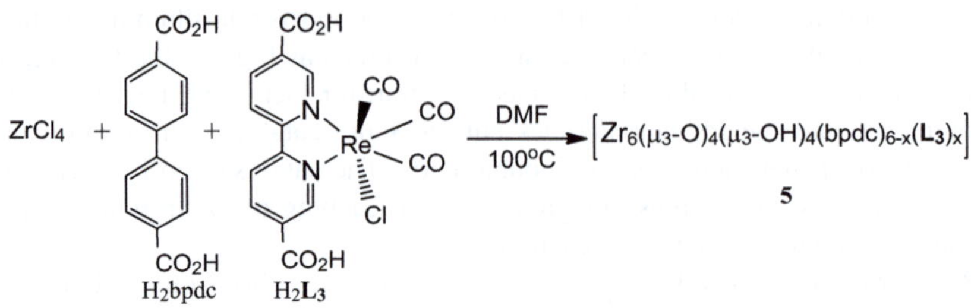

Scheme 1 Synthesis of **4** by doping [$Re^{I}(CO)_3$(5,5′,-dcbpy)Cl] into the UiO-67 framework

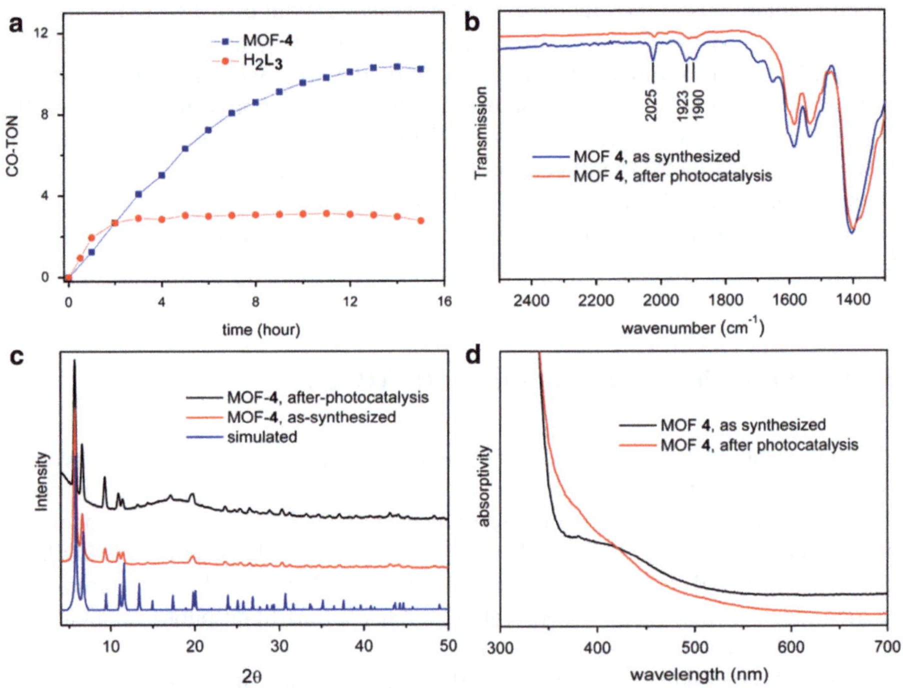

Fig. 4 (**a**) Plots of CO evolution turnover number versus time in the photocatalytic CO_2 reduction with **4** and the homogeneous control. (**b**) FT-IR of **4** before and after photocatalysis. (**c**) PXRD patterns of **4** before and after photocatalysis. The simulated PXRD for the UiO-67 is also shown. (**d**) UV–vis diffuse reflectance spectra of **4** before and after photocatalysis. Reprinted with permission from [35]. Copyright 2011 American Chemical Society

reaction. The recovered solid was found to be crystalline as shown by PXRD (Fig. 4c) but became inactive for CO generation. The loss of MLCT absorption band (Fig. 4d) and CO stretching vibrations (Fig. 4b) indicate that the detachment of Re-carbonyl moieties from the MOF backbones is responsible for the loss of catalytic activity. In addition, since the bimolecular pathway for CO_2 reduction is prohibited in **4** due to site isolation, the observed catalytic activity provides strong support for the

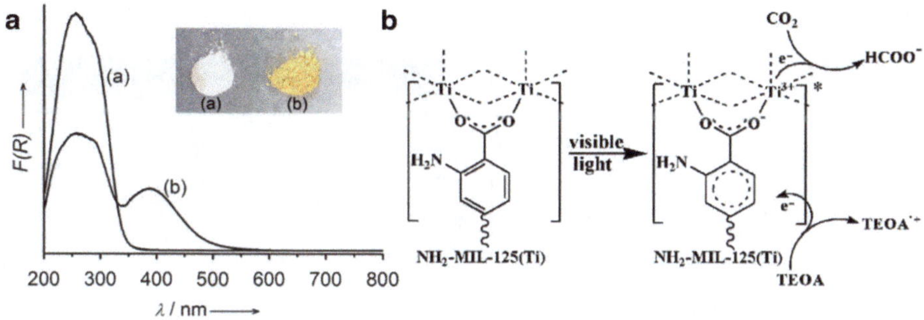

Fig. 5 Absorption spectra and proposed mechanism for photocatalytic CO_2 reduction of MIL-125(Ti) and NH_2-MIL-125(Ti). Reprinted with permission from [38]. Copyright 2012 Wiley-VCH

unimolecular pathway of photochemical CO_2 reduction. This work thus not only demonstrated the possibility of using MOFs for photocatalytic CO_2 reduction but also provided mechanistic insights for this reaction.

Li and coworkers reported in 2012 a photoactive MOF $Ti_8O_8(OH)_4(bdc\text{-}NH_2)_6$ (NH_2-MIL-125(Ti)) as the CO_2 reduction catalyst under visible light [38]. The MOF shows a visible absorption band that extends to 550 nm as a result of introducing the amino group (Fig. 5). In the photocatalytic reaction, CO_2 was reduced into HCO_2^- photochemically in acetonitrile with triethanolamine (TEOA) as the sacrificial reducing agent. It was proposed that, upon illumination, a long-lived charge transfer excited state was generated and reduced by TEOA to give a Ti^{3+} center which in turn reduced CO_2 to give HCO_2^-. NH_2-MIL-125(Ti) showed modest photocatalytic activity for photochemical CO_2 reduction, with a TON of 0.03 per Ti achieved in 10 h.

4 Photocatalytic Degradation of Organic Compounds with MOFs

Photocatalytic degradation of organic compounds is the most explored area in MOF photocatalysis [39–51]. **MOF-5**, with the framework composition of $Zn_4O(BDC)_6$ [52], was reported as a photocatalyst by Garcia et al. [39, 40]. **MOF-5** has an absorption spectrum that extends to 400 nm. It showed comparable activity for oxidative degradation of phenol as that of the conventional P-25 TiO_2 standard [40]. Interestingly, Garcia and coworkers found that **MOF-5** exhibits reverse shape selectivity when using different phenol substrates [39]. The more bulky phenol substrate, 2,6-di-*t*-butylphenol, exhibits a higher degradation rate compared to phenol. The reverse selectivity was explained based on the argument that unlike

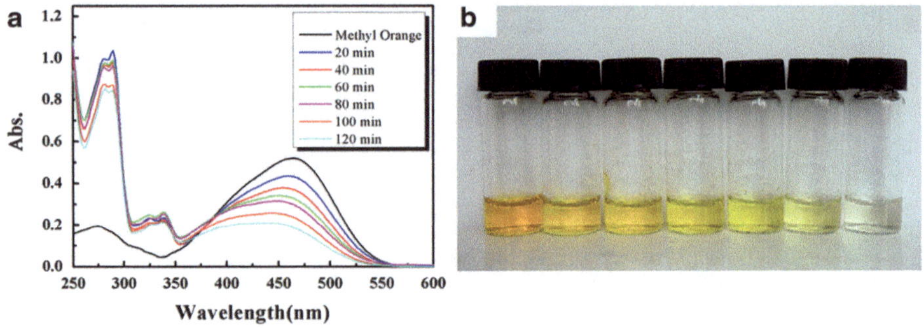

Fig. 6 UV-vis absorption spectra (**a**) and photograph (**b**) of methyl orange solution degraded by **UTSA-38** after the UV–vis light irradiation for different times. Reprinted with permission from [45]. Copyright 2011 Royal Society of Chemistry

smaller molecules which can diffuse into the MOF freely, large molecules remain on the external surface of the MOF where the rate of degradation is faster.

Photoactive MOFs have also been investigated for degradation of organic dyes such as methylene blue (MB) or rhodamine B (RhB). Ma et al. reported in 2012 a MOF constructed from two different ligands tetrakis[4-(carboxyphenyl)-oxamethyl]methane acid (H_4**L_4**) and tetrakis(imidazol-1-ylmethyl)methane (**L_5**) with the formula $[Co_2(\mathbf{L_4})(\mathbf{L_5})]\cdot 4.25H_2O$ (**5**) [49]. **5** has an absorption edge of 3.78 eV and exhibits catalytic activity for degradation of the dyes MB, RhB, and X3B. Under UV irradiation in the presence of **5**, MB degraded to 5.25% after 75 min, while RhB degraded to 34% and X3B degraded to 43% after 10 h, respectively. Moreover, **5** also catalyzed the degradation of MB under visible irradiation. After 5 h of visible irradiation, 49.6% of MB was degraded in the presence of **5**.

Chen et al. reported in 2011 the synthesis of a twofold interpenetrated MOF $Zn_4O(2,6\text{-naphthalenedicarboxylate})_3(DMF)_{1.5}(H_2O)_{0.5}\bullet 4DMF\bullet 7.5H_2O$ (**UTSA-38**) and its catalytic activity for degrading methyl orange (MO) [45]. **UTSA-38** has a band gap of 2.38 eV and exhibits catalytic activity under UV–visible light (Fig. 6).

Li et al. reported in 2012 a family of MOFs constructed from the ligands 1,2,4,5-benzenetetracarboxylic acid (btec) and 4,4′-bis(1-imidazolyl)biphenyl (bimb) and different metal ions for visible light-driven photocatalysis [47]. Degradation of the organic dye X3B was used to study the photocatalytic performance of the MOFs, and three of them, $[Co(btec)_{0.5}(bimb)]_n$ (**6**), $[Ni(btec)_{0.5}(bimb)(H_2O)]_n$ (**7**), and $[Cd(btec)_{0.5}(bimb)_{0.5}]_n$ (**8**), exhibited better performance than the commercial TiO_2 (Degussa P-25) under visible light (Fig. 7). The formation of hydroxyl radical OH· on the surface of the MOFs under visible light was detected by capture with terephthalic acid to form 2-hydroxyterephthalic acid. By measuring the photoluminescence (PL) intensity of the 2-hydroxyterephthalic acid, the authors were able to track the generation of hydroxyl radical and a zero-order kinetic was observed (Fig. 8).

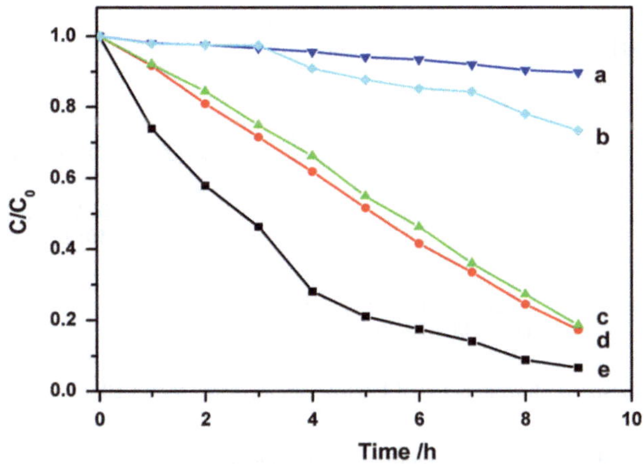

Fig. 7 Degradation profiles of X3B under visible light irradiation in the presence of (*a*) without catalyst, (*b*) Degussa P-25, (*c*) **6**, (*d*) **7**, and (*e*) **8**. Reprinted with permission from [47]. Copyright 2012 American Chemical Society

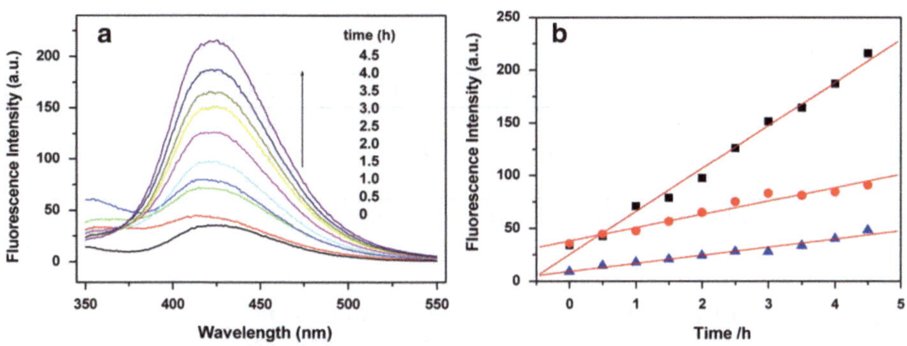

Fig. 8 (**a**) PL spectral changes observed during illumination of **8** in a 5×10^{-4} M basic solution of terephthalic acid (excitation at 315 nm). (**b**) Comparison of the induced PL intensity at 425 nm for MOFs **6** (*red*), **7** (*blue*), and **8** (*black*). Reprinted with permission from [47]. Copyright 2012 American Chemical Society

5 Photocatalytic Organic Transformations with MOFs

Compared to simple degradation of organic compounds, photocatalytic organic transformations often require better control of the reaction and better selectivity towards desired products, thus presenting a greater challenge for MOF photocatalysis. In the last few years, small molecule dyes such as $Ru(bpy)_3^{2+}$ and $Ir(ppy)_2(bpy)^+$ have been utilized as photoredox catalysts for various organic reactions [20–24]. Given the high cost of these noble metal-based

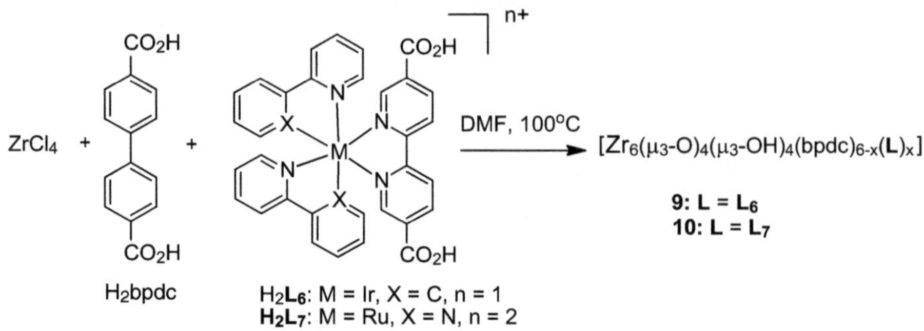

Scheme 2 Synthesis of phosphor-doped UiO-67 frameworks **9** and **10**

Table 1 Photocatalytic aza-Henry reactions with **9** and **10**

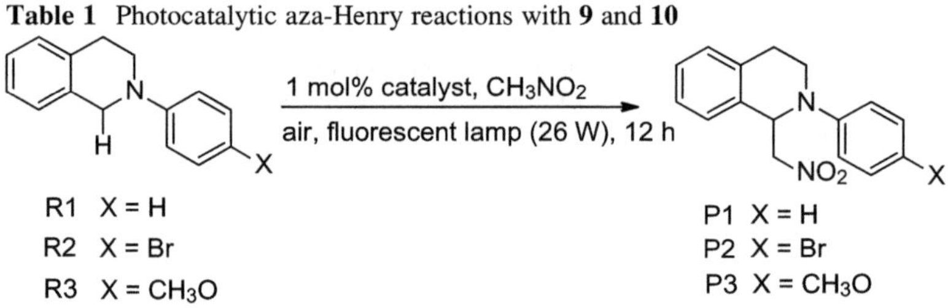

Compounds	Product	**9** (%)	**10** (%)	Et_2**L**$_6$[a] (%)	Et_2**L**$_7$[a] (%)
R1	P1	59	86	99	97
R2	P2	62	68	90	88
R3	P3	96	97	>99	>99

[a]Diethyl esters of **L**$_6$ and **L**$_7$ (Et_2**L**$_6$ and Et_2**L**$_7$) were used as homogeneous controls

Table 2 Photocatalytic aerobic amine coupling reaction based on **10**

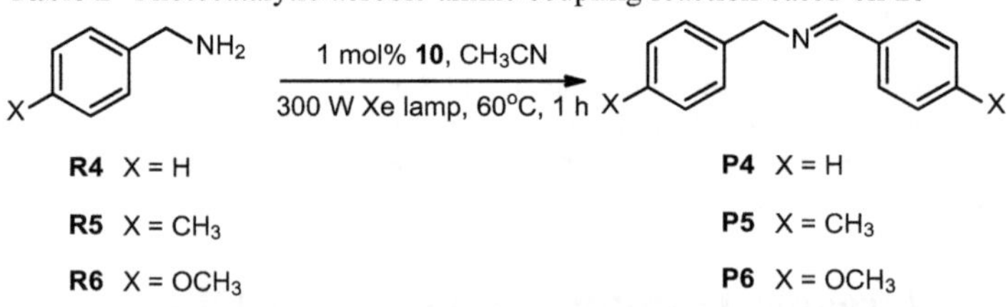

Compounds	Product	No light (%)	No catalyst (%)	Et_2**L**$_7$ (%)	**10** (%)	Reused **10** (%)
R4	P4	6	8	96	83	80
R5	P5				90	
R6	P6				46	

photocatalysts, it is desirable to develop heterogeneous version of these photocatalysts to facilitate recycling and reuse. In 2011, Lin and coworkers reported the preparation of highly stable photoactive UiO-67 MOFs **9** and **10**

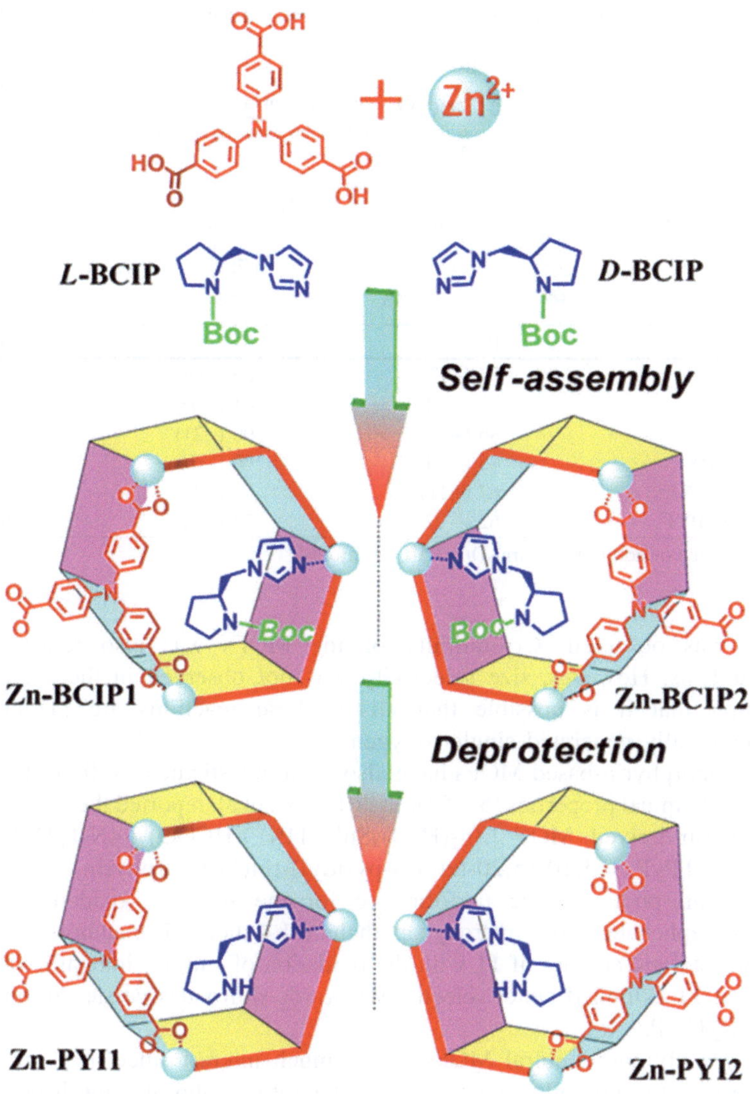

Fig. 9 Schematic representation of mirror image structures of **Zn-BCIP1** and **Zn-BCIP2** and their deprotected forms **Zn-PYI1** or **Zn-PYI2**. Reprinted with permission from [56]. Copyright 2012 American Chemical Society

doped with Ru or Ir-based phosphors (Scheme 2) [35]. They investigated aza-Henry reaction of tertiary amines, oxidative coupling of amines, and photooxidation of sulfides as the model reactions to illustrate the photocatalytic activities of the MOFs **9** and **10**. As shown in Table 1 and 2, **9** and **10** effectively catalyzed these reactions and gave slightly lower but comparable yields of the desiring product as homogeneous controls. The MOFs exhibited excellent reusability as no leaching of noble metals or decreasing of catalytic

Table 3 Conversions and the *ee* values (in the parenthesis) of the photocatalytic α-alkylation of aldehydes[a]

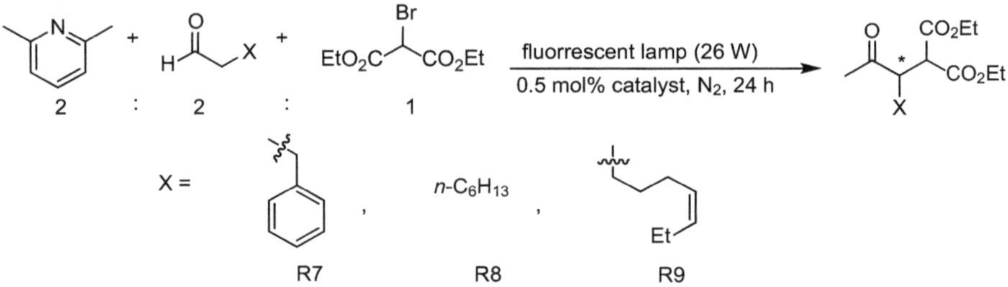

Catalyst	R7	R8	R9
Zn-**PYI1**	74 (+92)	65 (+86)	84 (+92)
Zn-**PYI2**	73 (−81)	61 (−78)	85 (−89)
Ho-TCA/L-PYI[a]	86 (+20)	90 (+21)	93 (+20)
Ho-TCA/D-PYI[a]	85 (−21)	90 (−20)	95 (−20)
MOF-150/L-PYI[a]	67 (+21)	78 (+24)	80 (+20)
MOF-150/D-PYI[a]	62 (−22)	73 (−22)	90 (−21)

[a]In the presence of additional chiral D/L-**PYI** (20 mol%)

activity was observed. Crystallinity of the MOFs was also retained after photocatalysis. However, size selectivity was not observed in these reactions, suggesting that it is possible that all of these reactions are mediated by photochemically generated singlet oxygen.

Metalloporphyrin-based MOFs have also been investigated for light-harvesting and photochemical properties [53, 54]. In 2011, Wu et al. reported the synthesis of a tin-porphyrin-derived MOF $[Zn_2(H_2O)_4Sn^{IV}(TPyP)(HCOO)_2]\cdot4NO_3\cdot DMF\cdot4H_2O$ (**11**, Sn^{IV}TPyP = 5,10,15,20-tetra(4-pyridyl)-tin(IV)-porphyrin) [55]. **11** showed high photocatalytic performance for the selective oxidation of 1,5-dihydroxynaphehalene to 5-hydroxynaphthalene-1,4-dione. The authors also tested the photocatalytic activity of **11** with the oxidation of sulfides into sulfoxides and found that **11** exhibited better selectivity towards sulfoxides versus sulfones than $Sn^{IV}(OH)_2$TPyP.

Catalytically active chiral MOFs are of much interest since they provide an effective way to immobilize chiral active sites into a solid, making it possible to prepare heterogeneous asymmetric catalysts. Combination of an asymmetric catalytic site and a photoactive functional component in the same MOF can potentially lead to highly efficient and stereoselective photocatalysis. Applying this strategy, Duan et al. reported in 2012 the incorporation of an asymmetric organocatalyst *L* or *D*-pyrrolidine-2-yl-imidazole (**PY1**) into a photoactive MOF with a triphenylamine moiety (Fig. 9) [56]. The resulting MOFs, **Zn-PYI1** or **Zn-PYI2**, were then used as heterogeneous catalysts for light-driven asymmetric α-alkylation of aliphatic aldehydes (Table 3). The photocatalytic reactions gave high yields and remarkable enantioselectivities. MOFs with different chiralities gave products with opposite selectivities. Control experiments using the Ho-TCA

(TCA = tricarboxytriphenylamine) MOF and MOF-150, which bear the same photoactive ligand, as photocatalyst with the same chiral adduct gave much lower ee value though comparable yields were achieved. The results indicated that the combination of the photosensitizer and the asymmetric organocatalyst in the same framework is the key to an enantioselective and photoactive MOF catalyst.

6 Conclusions and Outlook

Because photocatalytic systems typically require multiple components, MOFs are potentially advantageous in photocatalysis due to the ability to incorporate multiple functional groups in MOFs with well-defined spatial arrangements. In the past several years, MOFs have shown promising applications on photocatalysis with impressive results on photochemical hydrogen evolution, CO_2 reduction, and organic reactions. However, much more work is needed to prepare practically useful photocatalytic MOF or MOF-based assemblies.

Many MOFs tend to have modest hydrolytic stability which greatly limits their applications in solar energy utilization. Photoactive cross-linked polymers [57–59] could be an alternative choice for designing new photocatalytic systems. As MOFs and cross-linked polymers have shown great possibilities for multifunctionalization, further development in these areas can lead to novel materials with multicomponents and hierarchical structures for photocatalysis.

Acknowledgment We thank NSF (DMR-1308229) for funding support.

References

1. Fujishima A, Honda K (1972) Electrochemical photolysis of water at a semiconductor electrode. Nature 238:37–38
2. Youngblood WJ, Lee SA, Kobayashi Y, Hernandez-Pagan E, Hoertz PG, Moore TA, Moore AL, Gust D, Mallouk TE (2009) Photoassisted overall water splitting in a visible light-absorbing dye-sensitized photoelectrochemical cell. J Am Chem Soc 131:926–927
3. Reece SY, Hamel JA, Sung K, Jarvi TD, Esswein AJ, Pijpers JJH, Nocera DG (2011) Wireless solar water splitting using silicon-based semiconductors and earth-abundant catalysts. Science 334:645–648
4. Ohno T, Bai L, Hisatomi T, Maeda K, Domen K (2012) Photocatalytic water splitting using modified GaN:ZnO Solid solution under visible light: long-time operation and regeneration of activity. J Am Chem Soc 134:8254–8259
5. Wang X, Xu Q, Li M, Shen S, Wang X, Wang Y, Feng Z, Shi J, Han H, Li C (2012) Photocatalytic overall water splitting promoted by an phase junction on Ga_2O_3. Angew Chem Int Ed 51:13089–13092
6. Teets TS, Nocera DG (2011) Photocatalytic hydrogen production. Chem Commun 47:9268–9274

7. Du P, Eisenberg R (2012) Catalysts made of earth-abundant elements (Co, Ni, Fe) for water splitting: recent progress and future challenges. Energy Environ Sci 5:6012–6021
8. Lakadamyali F, Kato M, Muresan NM, Reisner E (2012) Selective reduction of aqueous protons to hydrogen with a synthetic cobaloxime catalyst in the presence of atmospheric oxygen. Angew Chem Int Ed 51:9381–9384
9. Luo S, Mejía E, Friedrich A, Pazidis A, Junge H, Surkus A, Jackstell R, Denurra S, Gladiali S, Lochbrunner S, Beller M (2013) Photocatalytic water reduction with copper-based photosensitizers: a noble-metal-free system. Angew Chem Int Ed 52:419–423
10. Hetterscheid DGH, Reek JNH (2012) Mononuclear water oxidation catalysts. Angew Chem Int Ed 51:9740–9747
11. Duan L, Bozoglian F, Mandal S, Stewart B, Privalov T, Llobet A, Sun L (2012) A molecular ruthenium catalyst with water-oxidation activity comparable to that of photosystem II. Nat Chem 4:418–423
12. Rosi NL, Eckert J, Eddaoudi M, Vodak DT, Kim J, O'Keeffe M, Yaghi OM (2003) Hydrogen storage in microporous metal-organic frameworks. Science 300:1127–1129
13. Murray LJ, Dinca M, Long JR (2009) Hydrogen storage in metal-organic frameworks. Chem Soc Rev 38:1294–1314
14. Lee J, Farha OK, Roberts J, Scheidt KA, Nguyen ST, Hupp JT (2009) Metal-organic framework materials as catalysts. Chem Soc Rev 38:1450–1459
15. Ma L, Abney C, Lin W (2009) Enantioselective catalysis with homochiral metal-organic frameworks. Chem Soc Rev 38:1248–1256
16. Della Rocca J, Liu D, Lin W (2011) Nanoscale metal-organic frameworks for biomedical imaging and drug delivery. Acc Chem Res 44:957–968
17. Horcajada P, Gref R, Baati T, Allan PK, Maurin G, Couvreur P, Férey G, Morris RE, Serre C (2012) Metal-organic frameworks in biomedicine. Chem Rev 112:1232–1268
18. Li J, Sculley J, Zhou H (2012) Metal-organic frameworks for separations. Chem Rev 112:869–932
19. Kreno LE, Leong K, Farha OK, Allendorf M, Van Duyne RP, Hupp JT (2012) Metal-organic framework materials as chemical sensors. Chem Rev 112:1105–1125
20. Nicewicz DA, MacMillan DWC (2008) Merging photoredox catalysis with organocatalysis: the direct asymmetric alkylation of aldehydes. Science 322:77–80
21. Nagib DA, Scott ME, MacMillan DWC (2009) Enantioselective α-trifluoromethylation of aldehydes via photoredox organocatalysis. J Am Chem Soc 131:10875–10877
22. Du J, Yoon TP (2009) Crossed intermolecular [2+2] cycloadditions of acyclic enones via visible light photocatalysis. J Am Chem Soc 131:14604–14605
23. Condie AG, González-Gómez JC, Stephenson CRJ (2010) Visible-light photoredox catalysis: Aza-Henry reactions via C–H functionalization. J Am Chem Soc 132:1464–1465
24. Lang X, Ji H, Chen C, Ma W, Zhao J (2011) Selective formation of imines by aerobic photocatalytic oxidation of amines on TiO_2. Angew Chem Int Ed 50:3934–3937
25. Kataoka Y, Sato K, Miyazaki Y, Masuda K, Tanaka H, Naito S, Mori W (2009) Photocatalytic hydrogen production from water using porous material $[Ru_2(p\text{-}BDC)_2]n$. Energy Environ Sci 2:397–400
26. Cavka JH, Jakobsen S, Olsbye U, Guillou N, Lamberti C, Bordiga S, Lillerud KP (2008) A new zirconium inorganic building brick forming metal organic frameworks with exceptional stability. J Am Chem Soc 130:13850–13851
27. Gomes Silva C, Luz I, Llabrés i Xamena FX, Corma A, García H (2010) Water stable Zr–benzenedicarboxylate metal-organic frameworks as photocatalysts for hydrogen generation. Chem Eur J 16:11133–11138
28. Wang C, deKrafft KE, Lin W (2012) Pt nanoparticles@photoactive metal-organic frameworks: efficient hydrogen evolution via synergistic photoexcitation and electron injection. J Am Chem Soc 134:7211–7214

29. Fateeva A, Chater PA, Ireland CP, Tahir AA, Khimyak YZ, Wiper PV, Darwent JR, Rosseinsky MJ (2012) A water-stable porphyrin-based metal-organic framework active for visible-light photocatalysis. Angew Chem Int Ed 51:7440–7444
30. Jordan P, Fromme P, Witt HT, Klukas O, Saenger W, Krausz N (2001) Three-dimensional structure of cyanobacterial photosystem I at 2.5 Å resolution. Nature 411:909–917
31. Hawecker J, Lehn J, Ziessel R (1983)Efficient photochemical reduction of CO_2 to CO by visible light irradiation of systems containing Re(bipy)(CO)3X or Ru(bipy)32+-Co_2+ combinations as homogeneous catalysts. J Chem Soc Chem Commun 0:536–538
32. Lin W, Frei H (2005) Photochemical CO_2 splitting by metal-to-metal charge-transfer excitation in mesoporous ZrCu(I)-MCM-41 silicate sieve. J Am Chem Soc 127:1610–1611
33. Liu Q, Zhou Y, Kou J, Chen X, Tian Z, Gao J, Yan S, Zou Z (2010) High-yield synthesis of ultralong and ultrathin Zn_2GeO_4 nanoribbons toward improved photocatalytic reduction of CO_2 into renewable hydrocarbon fuel. J Am Chem Soc 132:14385–14387
34. Yan S, Ouyang S, Gao J, Yang M, Feng J, Fan X, Wan L, Li Z, Ye J, Zhou Y, Zou Z (2010) A room-temperature reactive-template route to mesoporous $ZnGa_2O_4$ with improved photocatalytic activity in reduction of CO_2. Angew Chem Int Ed 49:6400–6404
35. Wang C, Xie Z, deKrafft KE, Lin W (2011) Doping metal-organic frameworks for water oxidation, carbon dioxide reduction, and organic photocatalysis. J Am Chem Soc 133:13445–13454
36. Koike K, Hori H, Ishizuka M, Westwell JR, Takeuchi K, Ibusuki T, Enjouji K, Konno H, Sakamoto K, Ishitani O (1997) Key process of the photocatalytic reduction of CO_2 using [Re(4,4′-X2-bipyridine)(CO)3PR3]+ (X = CH_3, H, CF_3; PR3 = Phosphorus Ligands): dark reaction of the one-electron-reduced complexes with CO_2. Organometallics 16:5724–5729
37. Takeda H, Koike K, Inoue H, Ishitani O (2008) Development of an efficient photocatalytic system for CO_2 reduction using rhenium(I) complexes based on mechanistic studies. J Am Chem Soc 130:2023–2031
38. Fu Y, Sun D, Chen Y, Huang R, Ding Z, Fu X, Li Z (2012) An amine-functionalized titanium metal-organic framework photocatalyst with visible-light-induced activity for CO_2 reduction. Angew Chem Int Ed 51:3364–3367
39. Llabrés IX, Corma A, Garcia H (2007) Applications for metal−organic frameworks (MOFs) as quantum dot semiconductors. J Phys Chem C 111:80–85
40. Alvaro M, Carbonell E, Ferrer B, Llabrési Xamena F, Garcia H (2007) Semiconductor behavior of a metal-organic framework (MOF) Chem Eur J 13:5106–5112
41. Wen L, Wang F, Feng J, Lv K, Wang C, Li D (2009) Structures, photoluminescence, and photocatalytic properties of six new metal−organic frameworks based on aromatic polycarboxylate acids and rigid imidazole-based synthons. Cryst Growth Des 9:3581–3589
42. Wang X, Wang Y, Liu G, Tian A, Zhang J, Lin H (2011) Novel inorganic-organic hybrids constructed from multinuclear copper cluster and Keggin polyanions: from 1D wave-like chain to 2D network. Dalton Trans 40:9299–9305
43. Wang F, Ke X, Zhao J, Deng K, Leng X, Tian Z, Wen L, Li D (2011) Six new metal-organic frameworks with multi-carboxylic acids and imidazole-based spacers: syntheses, structures and properties. Dalton Trans 40:11856–11865
44. Du J, Yuan Y, Sun J, Peng F, Jiang X, Qiu L, Xie A, Shen Y, Zhu J (2011) New photocatalysts based on MIL-53 metal-organic frameworks for the decolorization of methylene blue dye. J Hazard Mater 190:945–951
45. Das MC, Xu H, Wang Z, Srinivas G, Zhou W, Yue Y, Nesterov VN, Qian G, Chen B (2011) A Zn_4O-containing doubly interpenetrated porous metal-organic framework for photocatalytic decomposition of methyl orange. Chem Commun 47:11715–11717
46. Zhou S, Kong Z, Wang Q, Li C (2012) Synthesis, structure and photocatalytic property of a novel 3D (3,8)-connected metal-organic framework based on a flexible triphosphonate and a pentanuclear Cu(II) unit. Inorg Chem Commun 25:1–4

47. Wen L, Zhao J, Lv K, Wu Y, Deng K, Leng X, Li D (2012) Visible-light-driven photocatalysts of metal-organic frameworks derived from multi-carboxylic acid and imidazole-based spacer. Cryst Growth Des 12:1603–1612
48. Kan W, Liu B, Yang J, Liu Y, Ma J (2012) A series of highly connected metal-organic frameworks based on triangular ligands and d10 metals: syntheses, structures, photoluminescence, and photocatalysis. Cryst Growth Des 12:2288–2298
49. Guo J, Yang J, Liu Y, Ma J (2012) Two novel 3D metal-organic frameworks based on two tetrahedral ligands: syntheses, structures, photoluminescence and photocatalytic properties. CrystEngComm 14:6609–6617
50. Zhang Z, Yang J, Liu Y, Ma J (2013) Five polyoxometalate-based inorganic–organic hybrid compounds constructed by a multidentate N-donor ligand: syntheses, structures, electrochemistry, and photocatalysis properties. CrystEngComm 15:3843–3853
51. Du P, Yang Y, Yang J, Liu B, Ma J (2013) Syntheses, structures, photoluminescence, photocatalysis, and photoelectronic effects of 3D mixed high-connected metal-organic frameworks based on octanuclear and dodecanuclear secondary building units. Dalton Trans 42:1567–1580
52. Li H, Eddaoudi M, O'Keeffe M, Yaghi OM (1999) Design and synthesis of an exceptionally stable and highly porous metal-organic framework. Nature 402:276–279
53. Son H, Jin S, Patwardhan S, Wezenberg SJ, Jeong NC, So M, Wilmer CE, Sarjeant AA, Schatz GC, Snurr RQ, Farha OK, Wiederrecht GP, Hupp JT (2013) Light-harvesting and ultrafast energy migration in porphyrin-based metal-organic frameworks. J Am Chem Soc 135:862–869
54. Lee CY, Farha OK, Hong BJ, Sarjeant AA, Nguyen ST, Hupp JT (2011) Light-harvesting metal-organic frameworks (MOFs): efficient strut-to-strut energy transfer in bodipy and porphyrin-based MOFs. J Am Chem Soc 133:15858–15861
55. Xie M, Yang X, Zou C, Wu C (2011) A SnIV–porphyrin-based metal-organic framework for the selective photo-oxygenation of phenol and sulfides. Inorg Chem 50:5318–5320
56. Wu P, He C, Wang J, Peng X, Li X, An Y, Duan C (2012) Photoactive chiral metal-organic frameworks for light-driven asymmetric α-alkylation of aldehydes. J Am Chem Soc 134:14991–14999
57. Xie Z, Wang C, deKrafft KE, Lin W (2011) Highly stable and porous cross-linked polymers for efficient photocatalysis. J Am Chem Soc 133:2056–2059
58. Wang C, Xie Z, deKrafft KE, Lin W (2012) Light-harvesting cross-linked polymers for efficient heterogeneous photpcatalysis. ACS Appl Mater Interfaces 4:2288–2294
59. Wang J, Wang C, deKrafft KE, Lin W (2012) Cross-linked polymers with exceptionally high Ru(bipy)32+ loadings for efficient heterogeneous photocatalysis. ACS Catal 2:417–424

Struct Bond (2014) 157: 105–144
DOI: 10.1007/430_2013_121

Published online: 9 August 2013

Metal-Organic Frameworks for Photochemical Reactions

Raghavender Medishetty and Jagadese J. Vittal

Abstract In this chapter, various photochemical reactions involving coordination polymers and metal-organic frameworks have been described. These include structural transformations, post-synthetic modification of surfaces to create reactive sites, radicals which are not possible by convention synthetic routes, polymerization on the surfaces, and *cis–trans* isomerization of the guest molecules in the cavities using the dynamic behavior of the structures. Further, a few examples have been provided where [2+2] cycloaddition reaction has been used as a tool to monitor the changes in the solid-state structures in the absence of crystal structures. The structures of the resultant products of a number of polymerization reactions such as alkenoates were hampered due to the lack of suitable structural tool other than single crystal X-ray crystallography. On the other hand, most of the examples discussed in this chapter do maintain the single crystals at the end of the reactions making them amenable for structural elucidation to understand the reactivity completely.

Keywords 2+2 cycloaddition · Coordination polymers (CPs) · Crystal engineering · Cyclobutane derivatives · Ladder structures · MOFs · Photo-reactivity · Solid-state reactions · Structural transformation

Contents

R. Medishetty and J.J. Vittal (✉)
Department of Chemistry, National University of Singapore, Singapore 117543, Singapore
e-mail: chmjjv@nus.edu.sg

1 Introduction

Among the solid-state photochemical reactions, [2+2] cycloaddition reaction has been studied extensively in the past several decades [1–11]. The classical work of G. Schmidt and coworkers in 1960s paved away the field of solid-state organic photochemistry as well as crystal engineering [12–14]. Due to this historical importance, this double bonds containing spacer and terminal ligands have been incorporated into coordination polymers (CPs) and porous CPs (PCPs) which is also popularly called as metal-organic frameworks (MOFs) to design photo-reactive solids [15–17]. The metal-coordination bond offers very important role in directing the organic linkers and/or guest molecules to satisfy the geometry criteria required for the photoreactions. At the initial stage, enormous interests were shown to utilize various crystal engineering tools to design photo-reactive CPs and MOFs. Since the crystallization process is controlled by the kinetic factors, the challenge to design photo-reactive CPs still persists [18–20]. However, the interest is slowly shifting from design strategies to exploration and understanding the structure–function–property relationships. This includes molecular sensing, controlled guest release, pore size, and surface functionalization.

In this chapter, the following will be addressed: (1) How the olefin pairs in the spacer ligands as well as guest molecules can be aligned to undergo [2+2] cycloaddition reactions. (2) In the photochemical reactions, new σ bonds are formed between two carbon atoms to yield cyclobutane rings. This process brings structural transformations due to change in the structures, connectivity, dimensionality, or interpenetration. Hence, sections are mainly divided based on the changes in the dimensionality by the formation of cyclobutane rings from the olefin pairs. (3) The changes in the physical and chemical properties with the structural transformations will also be discussed. (4) Apart from solid-state [2+2] cycloaddition reactions, other photochemical reactions such as polymerizations and photodissociations have also been discussed. (5) Photochemical reactions in solution have also been briefly mentioned for the sake of completion.

2 Polymerization of Metal Complexes by [2+2] Cycloaddition Reactions

The first [2+2] photopolymerization reaction of a metal complex was reported for the mononuclear complex of uranyl chloride with *trans,trans*-dibenzylideneacetone (dba) shown in Fig. 1 resulted in a stereospecific *rctt* isomer of truxillic type 1,3-bis (3-phrnylprop-2-enol-oxo-2,4-bis(phenyl)cyclobutane) by Praetorius and Korn in 1910 [21, 22]. Later, Stobbe and Farber reported the formation of the similar dimer using $SnCl_4$ in 1925 [23] with a closely related ligand. To understand the reactivity, Alcock determined the crystal structure of $[(UO_2)Cl_2(dba)_2]$ (**1**) and $[(UO_2)Cl_2(dba)_2]{\cdot}2CH_3COOH$ (**2**), in which the uranium is present in octahedral coordination geometry surrounded by two Cl atoms, two O atoms of uranyl group, and two keto oxygen atoms from dba [24]. The ligand dba has *trans,trans* conformation, and the olefinic groups are separated by 4.09 and 3.51 Å for solvated and desolvated forms, respectively. Later, the complexes of Ni(II), Cu(II), and Zn(II) with a series of substituted cyclopentanones were found to undergo [2+2] photo-cycloaddition reactions [25] (Fig. 2).

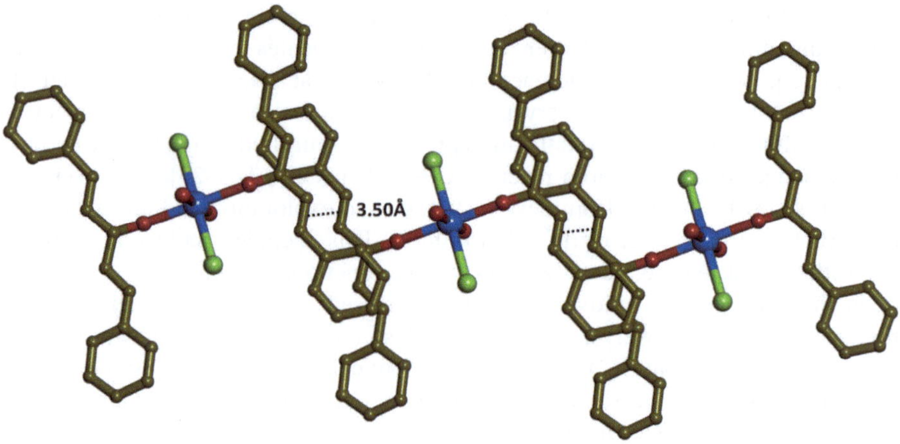

Fig. 1 The alignment of olefin groups of dba ligand in a uranyl chloride complex

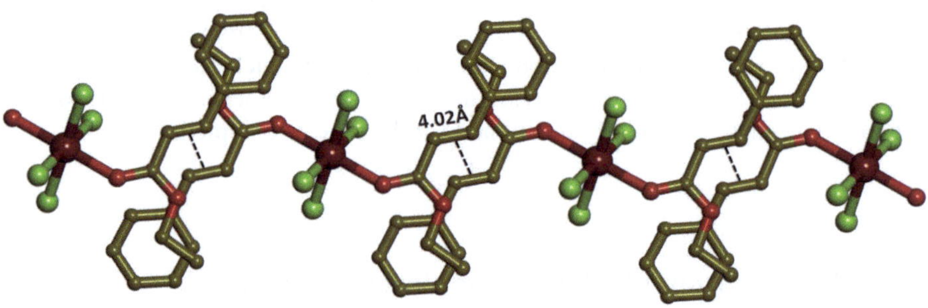

Fig. 2 The alignment of olefin groups of dba ligand in the $SnCl_4$ complex

2.1 Role of Ag· · ·C Interactions in Transferring a Dimer to a 1D CP

MacGillivray and coworkers showed that the intermolecular Ag· · ·C bond formed during [2+2] cycloaddition reaction in a discrete dinuclear Ag(I) complex can be used in the formation of an interesting 1D CP [26]. The 4spy ligands in the metal complex $[Ag_2(4spy)_4][TFA]_2$ (**3**) (4spy = 4-stryrylpyridine) have been paired up with the help of argentophilic and π–π interactions [26]. Though the C=C bonds in 4spy ligand pairs are aligned in a crisscross manner, the pedal motion of C=C bonds assisted the quantitative photo-reactivity of this compound and resulted in the formation of $[Ag_2($*rctt*-4-ppcb$)_2][(TFA)_2]$ (**4**) (*rcct*-4-ppcb = 1,3-bis(4′-pyridyl)-2,4-bis(phenyl)cyclobutane). This pedal motion was well studied by Ogawa and coworkers for stilbenes and azo-stilbenes [27]. Interestingly, this transformation

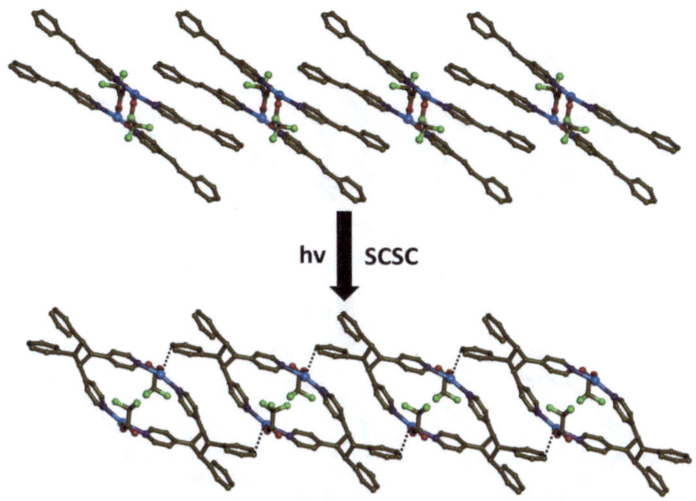

Fig. 3 Transformation of a metal complex into a 1D CP due to the formation of Ag· · ·C interactions

took place in an SCSC manner without loss of single crystallinity that enabled the structural elucidation of final dimerized compound by X-ray crystallography. The solid-state structure of the photoproduct showed that significant movements of the Ag(I) atoms and *rctt*-4-ppcb ligands took place during photoreaction which resulted in the formation of Ag· · ·C interaction (distance 2.64 Å) between the neighboring molecules giving raise to the formation of a 1D CP as shown in Fig. 3.

2.2 Polymerization of a Dimer Through [2+2] Cycloaddition Reaction

Recently another photoactive metal complex, $[Ag_2(Cl\text{-}spy)_4(ClO_3)_2]$ (**5**) (where Cl-spy = *trans*-1-(4-Cl-3-pyridyl)-2-(phenyl)ethylene), was synthesized with Ag(I) in tetrahedral coordination geometry and coordinated by N of two Cl-spy, O atoms of two ClO_3^- anions [28]. The solid-state structure analysis showed that only 50% of Cl-spy are aligned parallel between the dinuclear complexes and are separated by 3.73 Å as shown in Fig. 4. The other 50% are arranged beyond Schmidt's criteria (5.93 and 6.29 Å), and the two rings in the Cl-spy are twisted from coplanarity by 20.6°. As expected, on UV irradiation of this compound showed photo-reactivity of this compound in an SCSC manner and resulted in the formation of 1D CP.

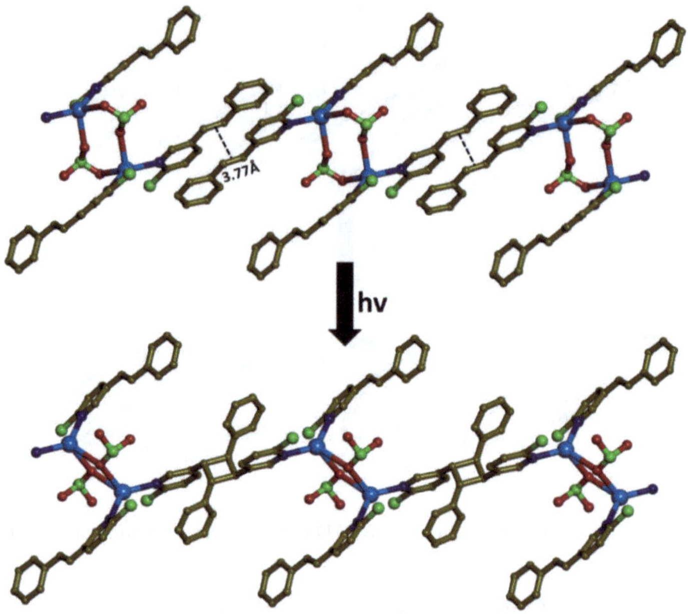

Fig. 4 Polymerization of a metal complex into a 1D CP through [2+2] cycloaddition reaction

2.3 Photo-Reactivity of 1D H-Bonded Zwitter-Ionic Metal Complexes

H-bonds are one of the strongest directional forces among all the supramolecular interactions. In a zwitter-ionic metal complex [Pb(bpe-H)$_2$(TFA)$_4$] (**6**) (where TFA = trifluoroacetate anion), complementary hydrogen-bonds between a coordinated monoprotonated bpe-H$^+$ cation and the neighboring oxygen atom of the TFA ligand assists the alignment of the bpe-H$^+$ pairs as shown in Fig. 5. The Pb(II) center is coordinated by four trifluoroacetates and two bpe-H$^+$ cations. The infinitely organized coordinated bpe-H$^+$ pairs easily undergo photocyclization under UV irradiation quantitatively to give the anticipated *rctt*-tpcb isomer [29]. The colorless block-like crystals turned into gel in 1 h and became powder after 25 h of UV irradiation. During photodimerization, two CF_3CO_2H were eliminated due to the proton transfer from bpe-H$^+$ to TFA ligand bonded to the Pb(II), resulted in [Pb(*rctt*-tpcb)(TFA)$_2$] (**7**). Since this is not an SCSC reaction, the structure of the photoproduct was determined after recrystallizing from methanol. Due to the reorganization of structures during crystallization, the connectivity of the crystallized product of [Pb(*rctt*-tpcb)(TFA)$_2$] is slightly different from predicted structure (shown in Fig. 5) based on the connectivity of the precursor.

A very similar zwitter-ionic complex, [Cd(bpe-H)$_2$(H$_2$O)(S$_2$O$_3$)$_2$] (**8**), was reported by Natarajan et al., in which two protonated [bpe-H]$^+$ ligands are coordinated to Cd(II) in *trans* fashion [30]. The pyridinium N–H proton at the other

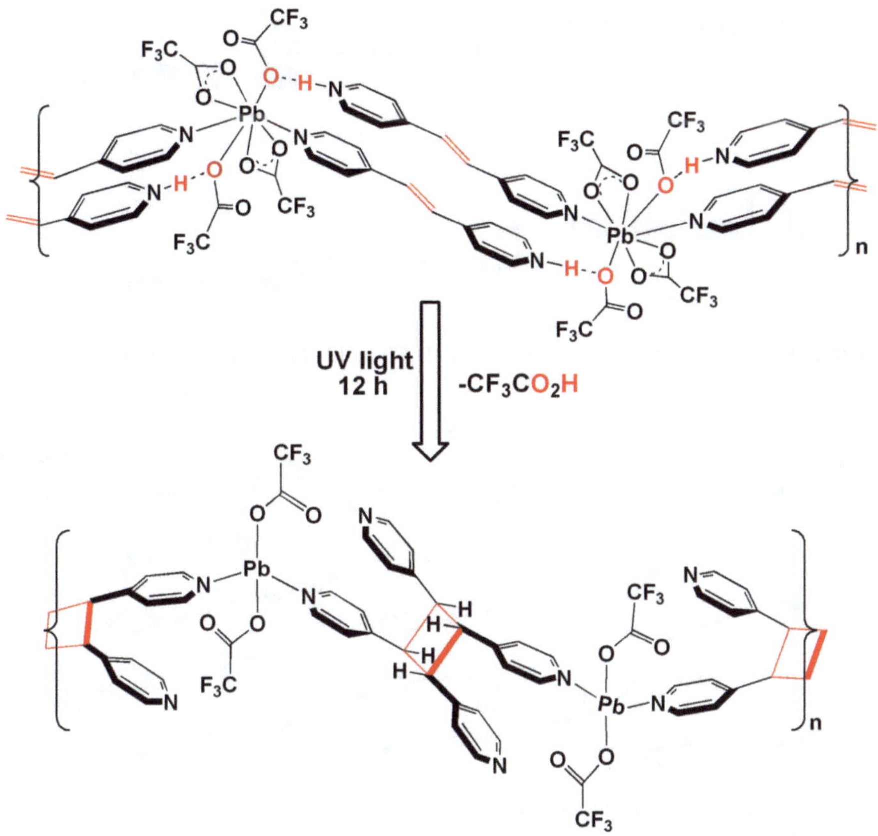

Fig. 5 A zigzag hydrogen-bonded zwitter-ionic Pb(II) complex undergoes polymerization reaction under UV light

end of $[bpe\text{-}H]^{+}$ is hydrogen bonded to the neighboring thiosulfate O atom to form a hydrogen-bonded 1D polymer. The olefinic groups are separated by 3.7 Å and upon UV irradiation of this compound showed quantitative [2+2] photo-cycloaddition as confirmed by ^{1}H-NMR and resulted in the formation of 1D chains as shown in Fig. 6. However, the solid-state structure of final compound was not characterized.

2.4 Polymerization of a Hydrogen-Bonded Metal Complex Containing Several Parallel and Crisscross C=C Bonds

An interesting H-bonded 1D metal complex, $[Zn(bpe)_2(H_2O)_4](NO_3)_2\cdot(8/3)H_2O\cdot(2/3)bpe$ (**9**), was reported, where the Zn(II) is present in octahedral geometry arising from the coordination of two bpe ligands in *trans* fashion and four water

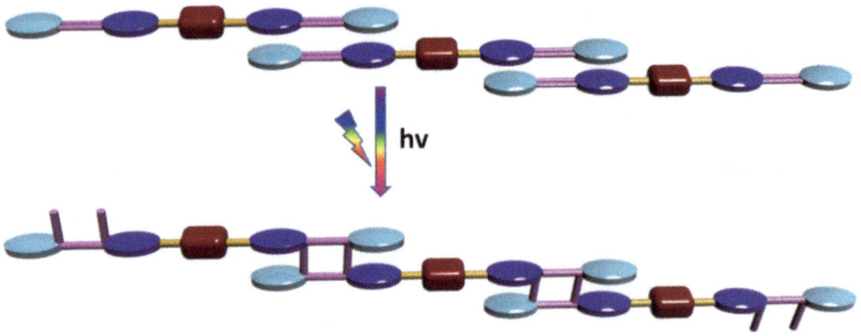

Fig. 6 Polymerization of a metal complex to a 1D CP by [2+2] cycloaddition reaction

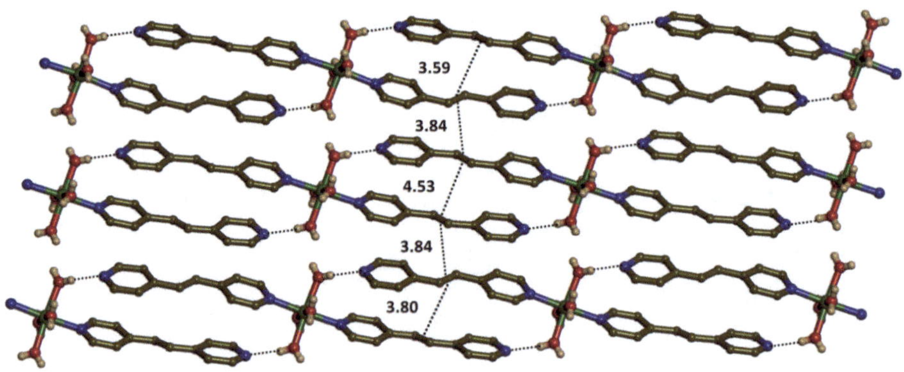

Fig. 7 A diagram showing the alignment of bpe molecules in the hydrogen-bonded metal complex, **9**

molecules in the equatorial sites [31]. A few guest bpe molecules were also found in the lattice along with lattice water. Only one side of the bpe is coordinated to Zn(II), and the other side of bpe is H-bonded to coordinated water and forming 1D H-bonded metal complex. These complementary H-bonds drive the bpe molecules to align parallel with each other. Interestingly, this compound consists of totally six different types of bpe molecules that are aligned in parallel as shown in Fig. 7. However, the C=C bonds are aligned both in parallel and crisscross manner and are separated 3.55–3.88 Å from each other. On UV irradiation of the single crystals for 25 h showed 46% yield tpcb comprising 39% *rctt* and 7% *rtct* isomers. But this percentage of *rtct* isomer formed is lower than expected percentage based on the number of antiparallel orientation of C=C bonds. This has been attributed to the partial pedal motion of the C=C bonds. This pedal motion in this compound has been found to be accelerated by the mechanical grinding by pestle and mortar for 5 min along with increased the photo-reactivity to 100% tpcb when exposed under UV light for 25 h. In the absence of the crystal structure, the resultant product is expected to be a 1D CP similar to those described in Sect. 2.3. Hence, this reaction can be considered as metal complex polymerization.

3 Photo-Reactivity in 1D Polymers

1D chains are the simplest of all the CPs. Of these, molecular ladder polymers dictated by the proximity of two or more metal ions, spacer ligands, and anions have attracted attention due to their interesting magnetic electronic and optical properties. Furthermore, among the 1D CPs only the ladder structure is useful to carry out [2+2] cycloaddition reaction. However, the synthesis of 1D photo-reactive CPs is often challenging, due to the formation of other kinetic products during the crystallization process.

3.1 1D to 1D′ Transformation

In 2005, Vittal's laboratory reported the synthesis of a ladder-like photo-reactive CP, $[\{(H_3CCO_2)(\mu\text{-}O_2CCH_3)Zn\}_2(\mu\text{-bpe})_2]$ (**10**), in which the Zn(II) atoms of the two linear Zn(bpe) strands are bridged by two acetate ligands and chelated by another acetate ligand. Here, the bridging acetate ligands act as "rungs" in the resultant ladder structure and control the distance between the two linear Zn(bpe) strands. As a result, the bpe ligands are aligned in parallel, and the olefinic groups are separated by 3.66 Å. UV irradiation of this compound showed quantitative photoreaction. However, the single crystals were not preserved in the photoproduct. In contrast, in another similar ladder compound $[\{(TFA)(\mu\text{-}O_2CCH_3)Zn\}_2(\mu\text{-bpe})_2]$ (**11**) in which half of the acetate groups were replaced by TFA, the single crystals were intact at the end of the UV experiment. Therefore, the structure of the photoproduct, $[\{(TFA)(\mu\text{-}O_2CCH_3)Zn\}_2(rctt\text{-tpcb})_2]$ (**12**) (*rctt*-tpcb = regio *cis*, *trans*, and *trans tetrakis*(4-pyridyl)cyclobutane), was able to be determined by X-ray crystallography (Fig. 8). During the photoreaction, the fluorine atoms might have assisted to withstand the strain in the single crystals during the structural transformation [32]. It is also worth noted that this is the first report on the [2+2] cycloaddition reaction observed in CPs in SCSC fashion.

These ladder polymers have also been obtained from mechanochemical synthesis by grinding the stoichiometric ratios of the precursors as shown in Fig. 9 [33].

Recently, it was reported that benzoate ligands can also be used as bridging ligands to bring a pair of linear [Pb(bpe)] linear strands to form photo-reactive ladder polymer [34]. In the compounds $[Pb_2(\mu\text{-bpe})_2(\mu\text{-}O_2C\text{–}C_6H_5)_2(O_2C\text{–}C_6H_5)_2]$ (**13**), $[Pb_2(\mu\text{-bpe})_2(\mu\text{-}O_2C\text{-}p\text{-Tol})_2(O_2C\text{-}p\text{-Tol})_2]\cdot 1.5H_2O$ (**14**), and $[Pb_2(\mu\text{-bpe})_2(\mu\text{-}O_2C\text{-}m\text{-Tol})_2(O_2C\text{-}m\text{-Tol})_2]$ (**15**), the bpe pairs have parallel orientations and found to undergo quantitative [2+2] cycloaddition reaction. Of these, the later compound with *m*-Tol ligand has been found to undergo SCSC transformation to $[Pb_2(rctt\text{-tpcb})(\mu\text{-}O_2C\text{-}m\text{-Tol})_2(O_2C\text{-}m\text{-Tol})_2]$ (**16**).

Lang and coworkers also succeeded in extending this design principle to synthesize a photo-reactive Cd(II) ID CP with ladder structure $[Cd_2(\mu\text{-bpe})_2(\mu\text{-}O_2C\text{–}C_6H_4\text{-}p\text{-Cl})_2(O_2C\text{–}C_6H_4\text{-}p\text{-Cl})_2]$ (**17**) using *p*-chlorobenzoate as the clipping

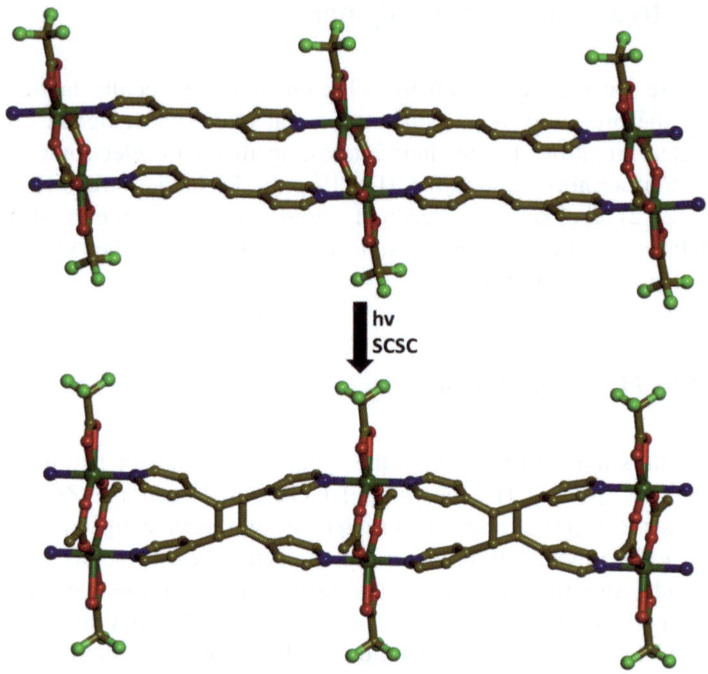

Fig. 8 1D to 1D′ structural transformation using an infinitely parallel bpe pairs

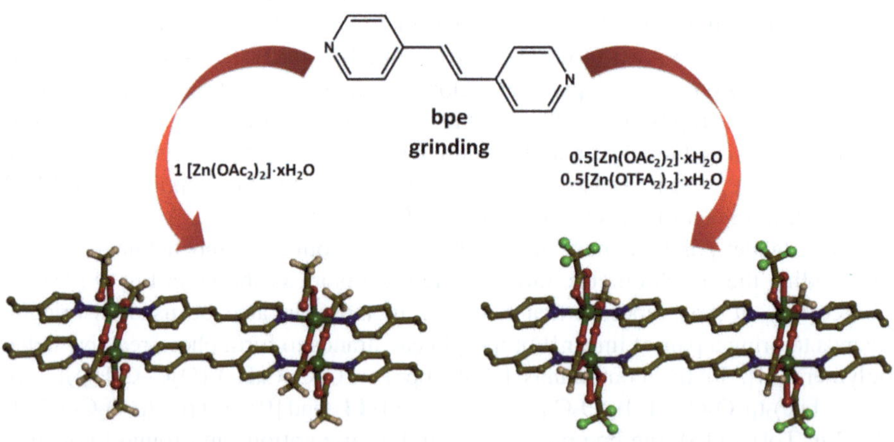

[{(H$_3$CCO$_2$)(μ-O$_2$CCH$_3$)Zn}$_2$(μ-bpe)$_2$] (10)

[{(TFA)(μ-O$_2$CCH$_3$)Zn}$_2$(μ-bpe)$_2$] (11)

Fig. 9 Schematic representation of mechanochemical synthesis of photo-reactive 1D CPs

agent for Cd(bpe) linear polymers. The quantitative photodimerization reaction was also accompanied by SCSC structural transformation. The Cd···Cd distance increased from 3.87 to 4.09 Å during the transformation as expected [35].

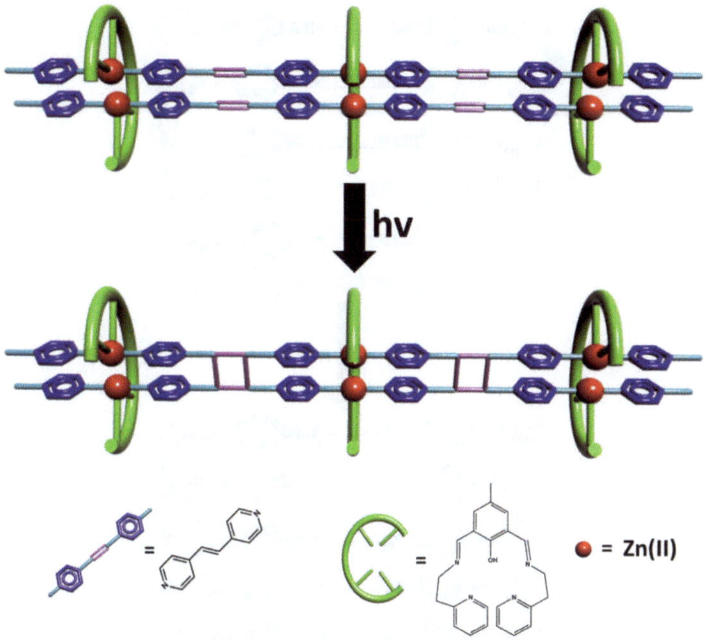

Fig. 10 Schematic representation of photoreaction of an infinitely aligned bpe pairs in a 1D Schiff base CP

3.2 Metal Template to Align the C=C Bonds

MacGillivray and coworkers have employed a template method to align a pair of bpe molecules using a dinuclear Zn(II) Schiff base complex in which the distance between the Zn(II) atoms is fixed at 3.1 Å. This Schiff base is able to coordinate and occupy all the equatorial positions of the two Zn(II) atoms in octahedral coordination geometry. Hence, the axial positions are freely available for the linear spacer ligands. First, they have shown that a pair of bpe can be aligned using this methodology. One of the two axial positions at the Zn(II) atoms was employed to synthesize a discrete tetranuclear compound which undergoes [2+2] cycloaddition reaction in an SCSC manner [36]. Later, they have extended the same template method successfully to construct a photo-reactive ladder structure. In this ladder structure, both the axial positions of the Zn(II) atoms were utilized, and as a result, the bpe pairs were aligned on both sides of the template as shown in Fig. 10. UV irradiation of this compound showed quantitative cycloaddition reaction resulting in another 1D ladder structure. However, single crystals could not survive after the photoreaction, probably attributed to the loss of water molecules during this photoreaction [37].

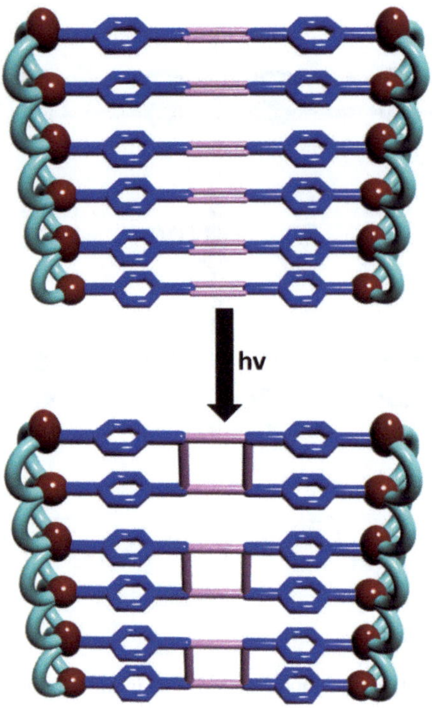

Fig. 11 Schematic representation of the structural transformation of **18–19** by photodimerization reaction

3.3 Alignment of the Spacer Ligand bpe in an Infinitely Parallel Fashion

Tong et al. in 2011 have shown that it is possible to align the bpe pairs one below the other infinitely in a 1D CP where the bpe ligands act as rungs [38]. Such rare ladder structure has been observed in $[Zn_2(1,2\text{-bdc})_2(bpe)]\cdot MeOH$ (**18**) (where 1,2-bdc is benzene-1,2-dicarboxylate). The nature of bonding of 1,2-bdc ligands dictates the intrachain Zn···Zn separations of 3.48 and 4.45 Å. One of the carboxylate group in one 1,2-bdc ligand chelates a Zn(II) in $\mu_2:\eta^1:\eta^1$ fashion, and the other carboxylate group bridges two more Zn(II) atoms in $\mu_2:\eta^0:\eta^2$ fashion in the 1D chain. On the other hand, each carboxylate group in the second 1,2-bdc bridges to two different Zn(II) atoms but overall bonded to three different Zn(II) atoms. Two adjacent 1D strands are bridged by bpe rungs to form this unique ladder structure as shown in Fig. 11. The nonbonding distances between the double bonds of the adjacent bpe ligands 3.84 and 3.96 Å satisfy Schmidt's topochemical criteria. Single crystal of this compound upon UV irradiation under a Hg lamp (400 W) for 5 h undergoes SCSC transformation to form $[Zn_2(1,2\text{-bdc})_2(tpcb)_{0.5}]$ (**19**) (tpcb = tetrakis(4′-pridyl)-cyclobutane). Interestingly, the solvent is lost in this structural conversion, and the interchain Zn···Zn distances changed to 3.57 and 4.43 Å [38].

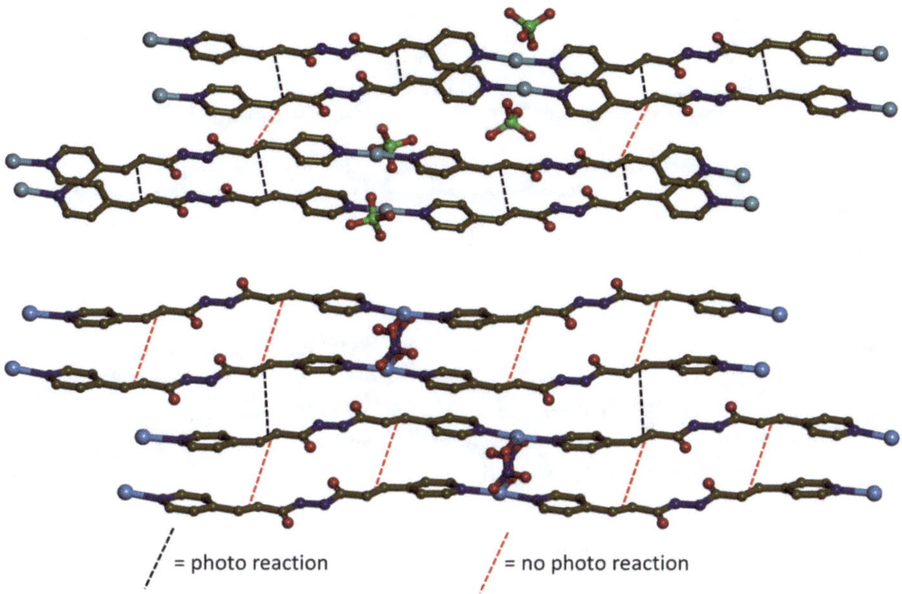

Fig. 12 The alignment of the photo-reactive olefin groups in 4PAH in Ag(I) 1D CPs

3.4 *Stepwise Double [2+2] Cycloaddition Reaction Assisted by Ag···Ag Interactions*

Biradha and coworkers employed Ag···Ag and Ag···π interactions to align the diolefinic organic ligand, *N*,*N*′-bis(3-(4-pyridyl)acryloyl)-hydrazine (4-PAH), and the counter anion also showed its role to control over the selective orientation of the olefinic groups for the photoreaction [39]. The 1D ladder CP, [Ag(4-PAH)(ClO_4)] (**20**), has parallel alignment of both the olefin groups, because of the Ag···Ag and π···π interactions supported by the perchlorate anion. The crystal packing showed the alignment of the olefin pairs in an asymmetric manner between the two linear strands, assisted by the Ag···π (olefin) interactions as shown in Fig. 12. Upon UV irradiation showed quantitative double [2+2] photoreaction between the two linear strands occurs by a stepwise mechanism as confirmed by the ^{1}H-NMR studies. Upon changing the counter anion to NO_3^-, similar double-stranded pairs were retained. But the interaction between the strands is so strong that the UV irradiation furnished only a mono-cyclobutane product.

3.5 *The Use of Conjugated Dienes for Double [2+2] Cycloaddition Reactions in 1D Polymers*

Bis-pyridyl ligands with conjugated dienes are indeed interesting since they can undergo double [2+2] cycloaddition reactions. Once such ligand was the bis-pyridyl

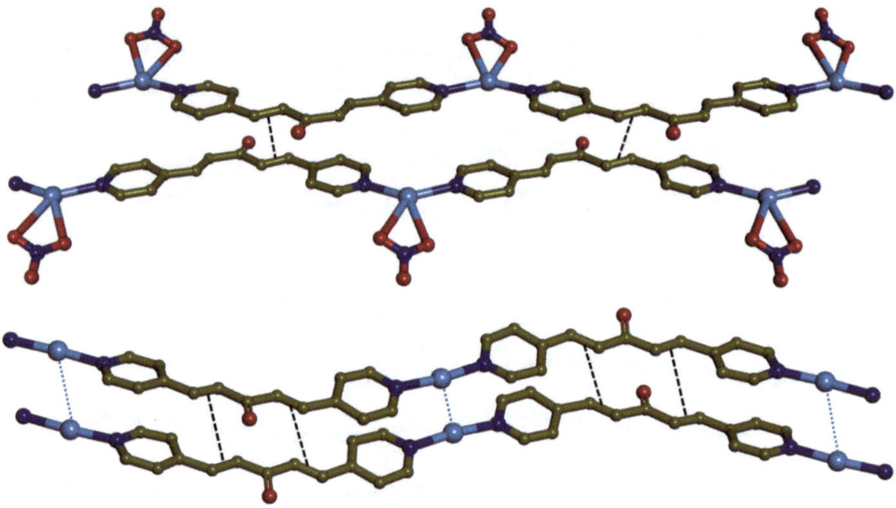

Fig. 13 The alignment of 1,3-pentadiene-3-one in Ag(I) 1D CPs

derivatives of 1,3-pentadiene-3-one used for making Ag(I) macromolecules and CPs by Biradha and coworkers [39]. The reaction of this ligand with $AgNO_3$ resulted in a 1D CP containing necklace-like loops made by Ag···Ag interactions as shown in Fig. 13. The ligand shows *cisoid–transoid* conformation, but only the C=C bonds with *cisoid* confirmation come closer between the interstrands. The other C=C bonds are "out of phase" with each other. Interestingly, this solid exhibits quantitative [2+2] photoreaction, but the nature of the product is unknown. It is likely that the product has 3D structure. However, changing the counter anion from NO_3^- to ClO_4^- showed completely different packing pattern due to *cisoid–cisoid* conformation of the ligand. The resultant linear 1D strands are paired up by Ag···Ag interactions to form ladder structure. This ladder compound is highly photo-reactive to yield 100% double [2+2] dimerized product as monitored by ^{1}H-NMR spectroscopy. Unfortunately, the structure of this 1D polymer was not confirmed [39].

3.6 Construction of Photo-Reactive 1D CP Using a Bioactive Dicarboxylates

Usually, d^{10} metal ions are used to construct photo-reactive CPs. Here is an unusual example of a transition metal ion to align a pair of olefin bonds in a biological active molecule. A heterocyclic diolefin chelidonic acid (H_2CDO) has been incorporated into a photo-reactive 1D ladder polymer $[Cu(CDO)(py)_2(H_2O)]$ (**21**) has been constructed from CDO anions with Cu(II) in presence of pyridine [40]. The neighboring six-membered pairs are aligned antiparallel with a distance of separation of 3.614 Å. Upon UV irradiation of this compound showed change in the color from blue to green without losing the single crystalline nature, and the single crystal

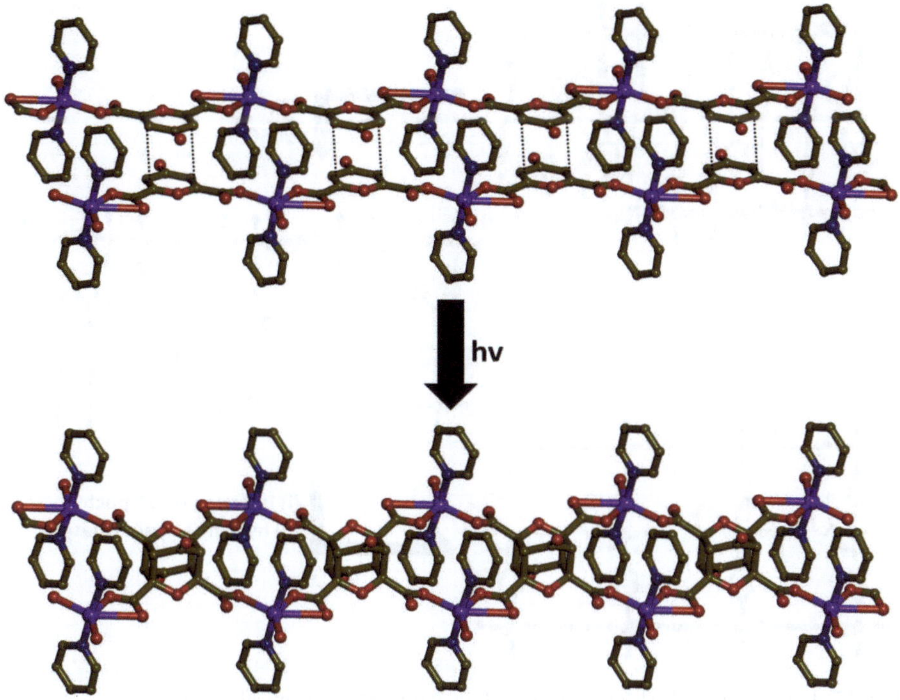

Fig. 14 Transformation of **21–22** by double photo-cycloaddition of a cyclic pairs

analysis of photo-irradiated crystal confirms the formation of photo-dimerized product [Cu(EE1)(py)$_2$(H_2O)] (**22**) and resulted in the formation of a similar 1D CP with cage-like ligand, EE1, and this ligand can be extracted and recrystallized as sodium salt. The structural transformation is shown in Fig. 14.

4 1D to 2D Transformation

4.1 Synthesis of Unique Furan-Based Ligand

Usually, spacer ligands have been made photo-reactive by incorporating C=C bonds in them. There are advantages as well as more challenges in making C=C containing terminal ligands to be photo-reactive. A simple ligand that can be used for this purpose is 4-styrylpyridine (4spy). The C=C double bonds can only be aligned in head-to-tail fashion between different polymeric strands, and hence 4spy can be aligned to undergo higher dimensional structural transformation using this versatile ligand. Metal complex polymerization (0D to 1D) was described earlier in Sect. 2.2. Here, an example for 1D to 2D structural transformation is described below.

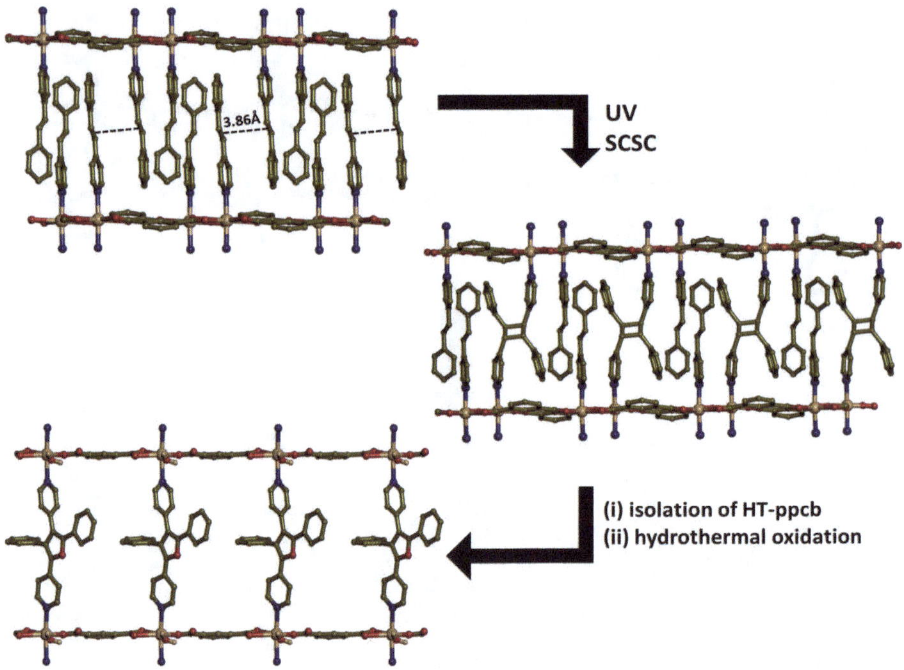

Fig. 15 Formation of 2D sheets from 1D CP by partial photochemical reaction

In the linear 1D CP [Cd(1,3-bdc)$(4spy)_2$] (**23**) (1,3-H_2bdc = 1,3-benzenedicarboxylic acid, 4spy = 4-styrylpyridine), the Cd(II) pairs are connected by four 1,3-bdc ligands as shown in Fig. 15 where one carboxylate bridges two Cd(II) centers and the other carboxylate ligand chelates a Cd(II) center in a plane. The two axial positions of the octahedral coordination environment of each Cd(II) atom are filled by the 4spy ligands, and the pyridyl pairs are aligned parallel. However, the olefinic bonds in the *head-to-head* pairs are misoriented and too far apart to undergo a photochemical [2+2] cycloaddition reaction. However, the 4spy ligands from neighboring chains interdigitate in a *head-to-tail* fashion. A closer look at the packing reveals that one of the two 4spy ligands is well aligned with its partner from the neighboring strand with C···C distance of 3.86 Å. In other words, 50% of the 4spy ligands can undergo HT dimerization above and below the 1D strands [26]. Indeed, SCSC reaction occurs under UV light, and the resultant compound [Cd(1,3-bdc)(4spy)$(\text{HT-ppcb})_{0.5}$] (**24**) (HT-ppcb = 1,3-bis(40-pyridyl)2,4-bis(phenyl)cyclobutane) has 2D sheet structure as shown in Fig. 15 [41].

The ligand HT-ppcb was isolated and combined with $CdCl_2 \cdot 2.5H_2O$, 1,3-H_2bdc, and dilute H_2O_2 under hydrothermal conditions yielded another interesting compound [Cd(H_2O)(1,3-bdc)(bpbpf)]·H_2O (**25**) (where bpbpf = 2,4-bis (4-pyridyl)-3,5-bis(phenyl)furan). During the hydrothermal reaction, the ligand underwent in situ oxidation reaction in the presence of oxygen to form this furan derivative [41].

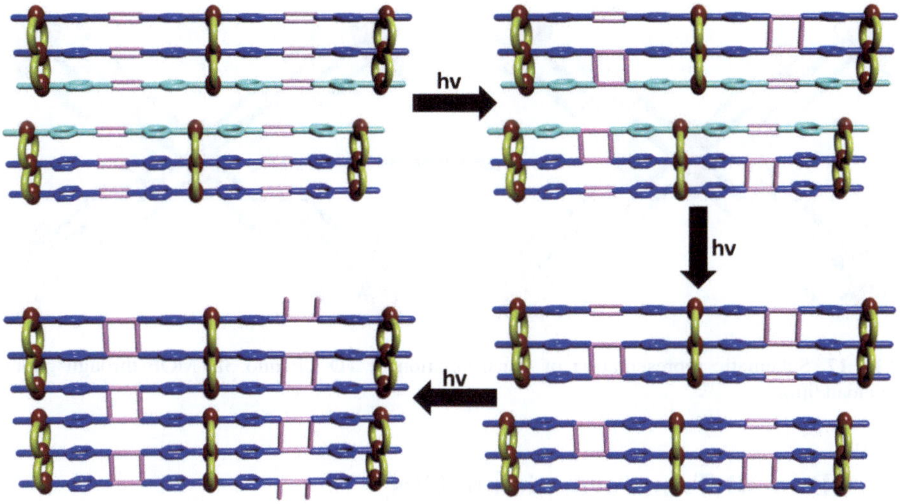

Fig. 16 Photo-reactivity of a triple-stranded 1D CP, **26**

4.2 Photo-Reactivity of a Triple-Stranded CP

Apart from the usual ladder structures formed by a pair of polymeric strands bridged by various anions as well as spacer ligands, one can also visualize three polymer strands to assemble to form an unusual ladder structure. This ladder can be considered as an intermediate in the self-assembly of 2D sheets. Such triple-stranded ladders are of interests to investigate the photo-reactivity, if C=C bonds in the three spacer ligands are aligned perfectly parallel within Schmidt's criteria.

Indeed, a photo-reactive triple-stranded ladder, $[Pb_3(bpe)_3(O_2C–CF_3)_4\ (O_2CCH_3)_2]$ (**26**), containing parallel alignment of the three bpe spacer ligands has been stumbled upon by Vittal and coworkers [42]. In this compound, Pb(II) in the center strand is bridged by an acetate and a trifluoroacetate ligand, and the terminal Pb(II) atoms of this 1D ribbon are chelated by trifluoroacetate anions, which prevent the further assembly to form 2D CP. This compound showed interesting photo-reactivity. UV irradiation of these single crystals within 10 h showed the expected 66% of dimerization of bpe due to the photoreaction between any two bpe molecules across the strands and left the third one unreacted as shown in the time versus percentage conversion plot as monitored by ^{1}H-NMR studies in solution. Surprisingly, further photo-irradiation showed quantitative photoreaction of bpe. Moreover, grinding of the single crystals to powder form accelerates the second step to completion in 40 h [42]. The authors suggested that there might be cooperative movements of these triple strands during the process which aligned the unreacted bpe between the intermolecular strands and thus the observed quantitative [2+2] cycloaddition as shown in Fig. 16. Such molecular movements have been noted before [43, 44].

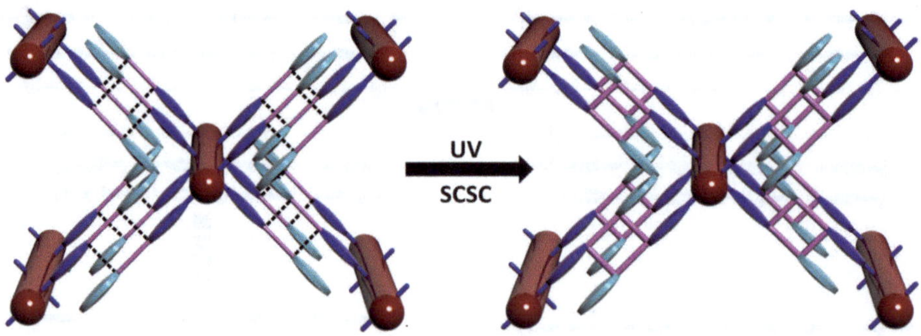

Fig. 17 Schematic representation of transformation of 1D CP into 3D MOF through photo-cycloaddition

5 1D to 3D Transformation in CPs

5.1 An Unusual Orientation of bpe Pairs for Photo-Reactivity

Tong and coworkers synthesized two photo-reactive 1D CPs, [M(1,2-chdc)(bpe)$_2$ (H_2O)$_2$]·H_2O (M = Zn(II) (**27**), Mn(II) (**28**), 1,2-chdc = *trans*-1,2-cyclohexanedicarboxylate) [45]. In these CPs, M(II) is present in octahedral coordination geometry, which has been coordinated by two carboxylate O of chdc; two more sites are coordinated by N of bpe and the coordination sphere is completed by water molecules. The connectivity of the chdc ligands with two different metals leads in the formation of 1D chain, and the bpe molecules are interestingly coordinated from only one N, and the other side is H-bonded to the aqua ligands in the neighboring chain. Similarly, the coordinated bpe in the next chain is, in turn, also H-bonded to the first chain. Such complementary hydrogen bonds align the double bonds of these two bpe in parallel with a C···C separation of 3.27 Å as shown in Fig. 17. Upon UV irradiation, the C=C bonds from neighboring bpe molecules form cyclobutane rings quantitatively without the loss of single crystallinity. The new bonds formed in the direction normal to the propagation of 1D CP transformed product into 3D network structure with NbO topology [45].

6 Photo-Reactivity in 2D CPs

6.1 Photo-Reactivity of a Layer CP yielding an unusual product

Here, we describe the photo-reactivity in an unusual crisscross alignment of the reactive C=C bonds in adjacent CP layers [46]. In the 2D layer compound [Cd$_2$(fumarate)$_2$(H_2O)$_4$] (**29**), the building block comprises a Cd(II) dimer. Each

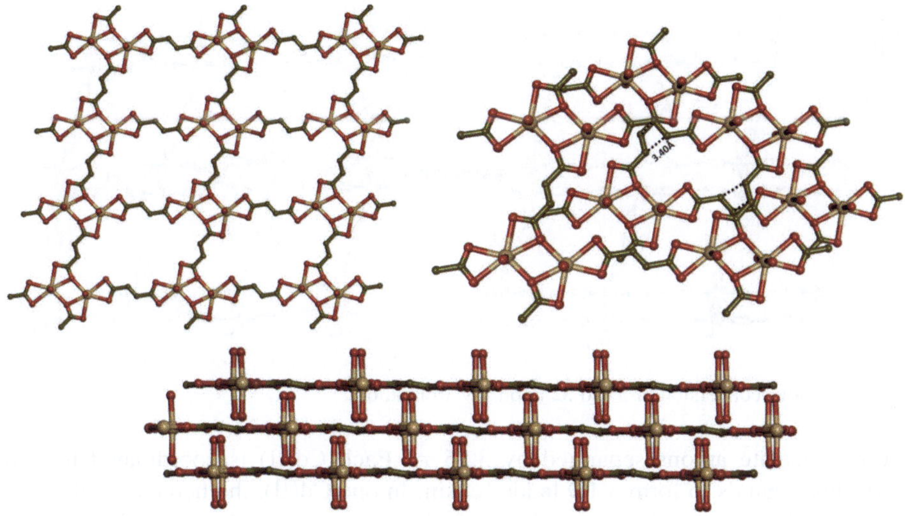

Fig. 18 Photo-reactivity of fumarates in 2D layered CP

Cd(II) is bonded to a chelating fumarate ion and two bridging fumarate anions and forms a Cd_2O_2 node as shown in Fig. 18. Further connectivity of $Cd_2(fumarate)_{4/2}$ dimer produced the expected (4,4) sheet structure. Each Cd(II) is further bonded to two aqua ligands in a *trans* manner. These coordinated water molecules are extensively hydrogen bonded to the adjacent layers. The staggered arrangements of the adjacent layers in ABAB··· fashion left no voids in the structure. On the contrary, the olefinic groups of the fumarates in the adjacent layers are separated by a distance of 3.38 Å in a crisscross fashion. Usually, such arrangement will undergo pedal-like motion in the organic molecules and 1D CPs but highly improbable in 2D layers. Hence, this cycloaddition reaction is to yield *rtct* isomer. Interestingly, UV irradiation experiments yielded the expected unusual *rtct*-cyclobutane-1,2,3,4-tetracarboxylate in the quantitative yield as monitored by the appearance of the singlet at 3.31 ppm due to the cyclobutane C–H proton in the ^{1}H-NMR spectrum of the dissolved product. Due to ABAB··· type of packing, cyclobutane ring formation can yield either bilayer structure, if the reaction occurs exclusively between two layers, or 3D structure if C–C bonds are formed randomly on both sides of the layer. Since the crystals crashed during UV experiment, this interesting information is not revealed [46].

6.2 2D to 2D′ Transformation

Miao and Zhu have synthesized a photo-reactive twofold interpenetrated 2D layered structure of $[Cd_2(CH_3COO)_2(3\text{-sb})(bpe)_{2.5}(H_2O)]\cdot 4H_2O$ (**30**) (3-sb = 3-sulfonatobenzoate) [47]. In the building block, two Cd(II) atoms are bridged

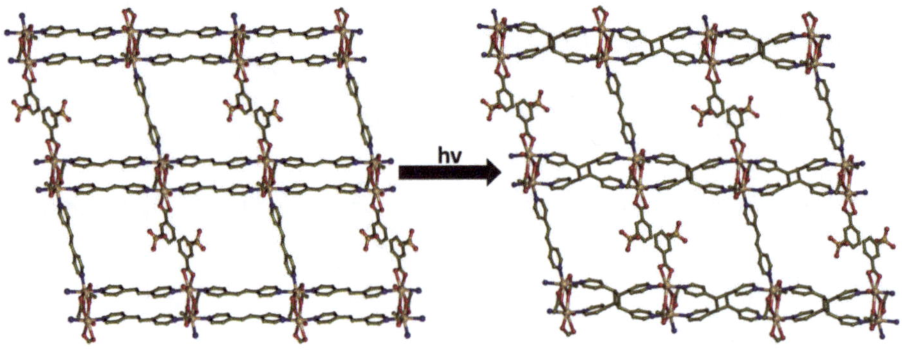

Fig. 19 Photo-conversion of **31** to **32** upon UV irradiation

by two acetate anions separated by 3.85 Å. Each Cd(II) is coordinated to two *trans* bpe ligands to form a 1D ladder chain. In one Cd(II), the hepta coordination geometry is completed by a chelating 3-sb anion and an aqua ligand. Both the acetate anions not only bridge the Cd(II) atoms but also chelate the second Cd(II) atom. The equatorial position in the pentagonal bipyramid geometry of this Cd(II) is completed by another bpe ligand bridging the neighboring ladder chains. This leads to the formation of a brick wall type 2D structure, and the voids created by this topology are minimized by twofold interpenetration. It is noted that the $[Cd_2(OAc)_2]$ unit dictates infinite parallel alignment of the bpe pairs and makes this crystal photoactive. The ^{1}H-NMR spectrum of the UV irradiated product showed that all these aligned bpe molecules reacted. The product $[Cd_2(CH_3COO)_2(3\text{-sb})(bpe)_{0.5}(4,4'\text{-tpcb})_2(H_2O)]\cdot 4H_2O$ (**31**) was independently synthesized by exposing the reaction mixture of **1** in solution under sunlight for several hours and then slowly evaporated at room temperature. The structure determination by X-ray crystallography reveals the retention of interpenetrated 2D layer structures. Another interesting aspect of this reaction is that **31** was obtained only with a high-power mercury lamp (300 W, $\lambda = 367$ nm) both in solution and the solid state after irradiation of 8 h but not using a low power UV lamp (30 W) with $\lambda = 253.7$ nm for 9 days [47] (Fig. 19).

6.3 2D to 2D′ Transformation and Its Thermal Cyclo-Reversion

The cleavage of the cyclobutane rings to the corresponding olefins is of interest since reversible photochemical reaction can be investigated for potential applications in optical storage and sensing. However, such cleavage of cyclobutane rings in CPs has never been observed. Recently, Vittal's laboratory has succeeded in cleaving the cyclobutane rings once formed in a 2D CP by thermal treatment [48].

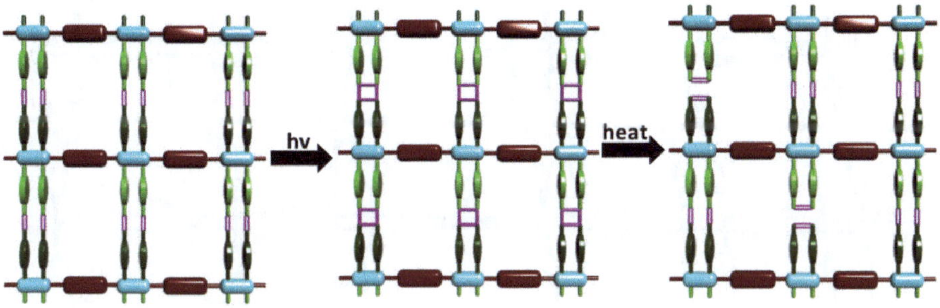

Fig. 20 Schematic representation of 2D to 2D photo-transformation of **33** and its thermal reversibility

The following photo-reactive 2D CPs have been synthesized using an asymmetric ligand, *trans*-2-(4′-pyridyl)vinylbenzoic acid (Hpvba): $[Cd_2(pvba)_2(tbdc)(dmf)_2]$ (**32**), $[Co_2(pvba)_2(tbdc)\ (dmf)_2(H_2O)_2]$ (**33**), and $[Ni_2(pvba)_2(tbdc)(dmf)_2(H_2O)_2]$ (**34**), where tpdc = 2,3,5,6-tetra-bromobenzene dicarboxylate. In the 2D sheet, $[Cd_2(O_2C–C)_4(pyridyl)_2]$ core is the repeating unit connecting a pair of pvba ligands which are aligned in *head-to-tail* parallel fashion in the orthogonal direction and the tbdc ligands in the perpendicular direction. The C=C bonds in these pvba ligand pairs are well aligned to undergo [2+2] cycloaddition reaction quantitatively in the solid state under UV light, yielding the cyclobutane derivative. This photochemical reaction has been accompanied by a solid-state structural transformation from one 2D structure to another as shown in Fig. 20 [48].

Further, the cyclobutane rings in the photodimerized product of **32** can be cleaved by heating to 220°C for 12 h under vacuum in two different ways to yield a mixture of *trans* and *cis* isomers of pvba quantitatively as monitored by the ^{1}H-NMR spectroscopy. This transformation as depicted in Fig. 20 appears to be the first attempt to reverse the cyclized product in CPs. From the integration of ^{1}H-NMR spectrum, the ratio between the *cis* and *trans* isomers has been observed to be 66.5% and 33.5%, respectively. Formation of other isomer of pvba has also been supported by a new peak observed in photo luminescence at 370 nm, which is different from *trans* pvba and the dimer. Under similar conditions, the photoproducts of **33** and **34** showed scission of cyclobutanes up to 56% and 67%, respectively. Similar to **32**, the new peaks also have been observed during the scission of cyclobutane, which are related to *cis* isomer of pvba. However, the intensities of these peaks are very small for integration from ^{1}H-NMR spectral data.

6.4 Influence of 2D to 2D′ Structural Conversion on the Gas Sorption Properties

Pore surface modification is the topic of current interest due to the ease in tuning the properties for different applications. Kitagawa and coworkers have used [2+2] cycloaddition reaction as a post-synthetic method to modify the pore structure

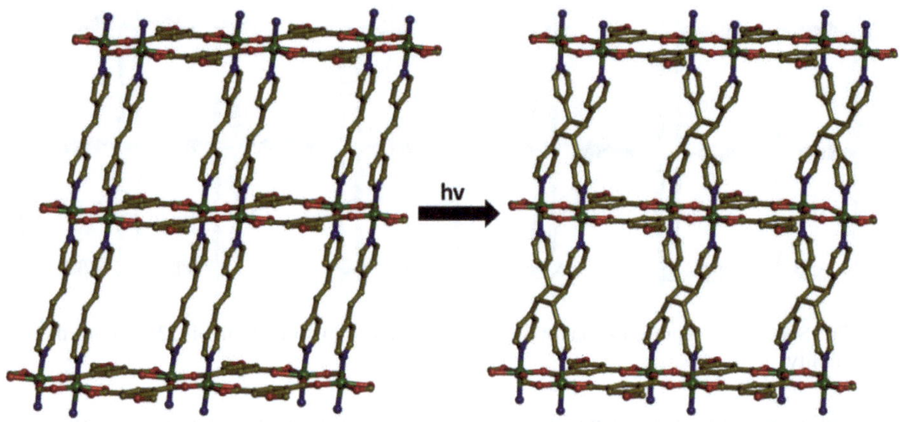

Fig. 21 The transformation of pores upon [2+2] photo-cycloaddition of bpe in **35**

and surface area in a 2D sheet structure [49]. They synthesized a photo-reactive twofold interpenetrating 2D CP, $[Zn_2(moip)_2(bpe)_2(DMF)_2]$ (**35**) (H_2moip = 5-methoxy-isophthalic acid). The 1D polymer formed by $Zn_2(moip)_2$ was further linked by pairs of bpe ligands. These bpe ligands are oriented in parallel, and the olefin groups are separated by 3.71 Å. In the resultant 3D structure obtained by twofold interpenetration of these sheets, 1D channels are created with a cross section of 3.2 Å × 4.5 Å with the solvent-accessible space of 27.0% of the unit cell volume. Upon UV irradiation of this compound, showed quantitative photo-cycloaddition in an SCSC manner to form $[Zn_2(moip)_2($*rctt*-tpcb$)(DMF)_2]$ (**36**). During this process, the size of the 1D channel has been enlarged due to the cyclobutane formation as shown in Fig. 21. Different gas sorption studies have been performed, which showed no uptake of N_2 at 77 K, which might be due to slow diffusion rate. However, these compounds showed better adsorption of CO_2 at higher temperature, 195 K. The activated precursor sample showed three stepwise adsorptions with the initial uptake up to 0.06 kPa and followed by a steep rise from 0.06 to 1 kPa and then gradual pore filling at further increase in pressure of CO_2 and also showed a hysteresis at the step 0.06 kPa. In contrast, the activated photoproduct showed normal type I isotherms without any discontinuous steps. This observation attributed the differences in the pore shape and the flexibility of the framework. Hence, one could engineer the pore size and surface through photo-reactivity [49].

6.5 2D to 2D′ and 3D to 3D′ Structural Transformations

During the synthesis of 3D MOF, Natarajan's laboratory separated an interesting 2D photo-reactive MOF, $[Cd_2(bpe)_3(H_2O)_4(S_2O_3)_2]$ (**37**), as a kinetic product [30]. In this solid-state structure, 1D chains have been formed between Cd(II) and S (in μ_2 mode) from thiosulfate, and the alternating Cd(II) atoms are coordinated

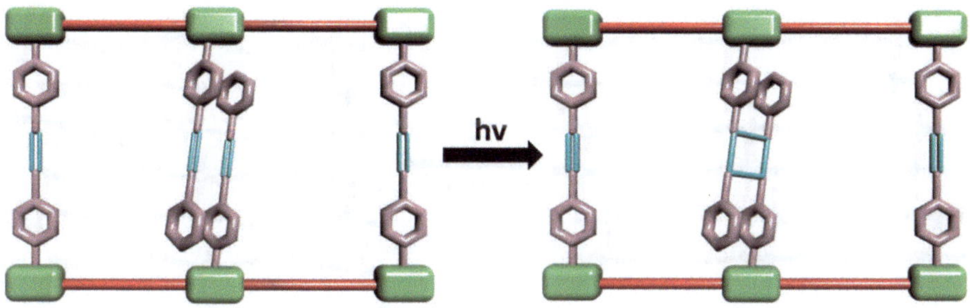

Fig. 22 A diagram showing the photo-reactivity between the dangling bpe molecules in **37**

by two types of bpe ligands. Among the first type is cross-linking the "Cd(II)-S" chains, and the other type is coordinated to Cd(II) and hangs from other side without coordinating. These "pendant" bpe from the adjacent "Cd(II)-S" chains is perfectly aligned parallel, and the olefinic groups are separated by a distance of 4.02 Å. Upon UV irradiation of this compound showed complete photo-reactivity of these aligned bpe molecules as shown in Fig. 22, which accounts for 67% of whole bpe molecules in the compound. This reaction is expected to generate another 2D layer structure through the formation of cyclobutane rings.

Similarly, thermodynamic product [Cd(bpe)S_2O_3] (**38**) was obtained by heating the reaction solution of **38** at 60°C for 72 h. The structure is built from a tetranuclear $Cd_4(S_2O_3)_4$ SBU made by Cd_4S_4 ring. The other two coordination sites in each tetrahedral Cd(II) are occupied by bpe ligands. In other words, there are four pairs of bpe ligands tetrahedrally oriented in this tetranuclear cluster to produce fourfold interpenetrated diamondoid structure. Each bpe pair from two sulfur bridged Cd(II) atoms is aligned in parallel but the olefinic groups are oriented in antiparallel manner in two bpe pairs and separated by 3.79 Å. Upon UV irradiation of this compound showed up to ~80% photo-reactivity and giving only *rctt* isomer due to pedal motion of the C=C bonds [30].

6.6 2D to 3D Transformation

A unique 2D bilayered CP, $[Mn_2L_2(H_2O)_2]\cdot 3H_2O$ (**39**) (L = E-5-(2-(pyridin-4-yl) vinyl)isophthalic acid ligand), was reported by Wu and coworkers [50]. Each Mn(II) atom with distorted octahedral coordination geometry is coordinated to a chelated carboxylate O atoms, two *syn* and *anti* bridging O atoms from two different ligands, one N atom from fourth ligand, and a water molecule. The Mn(II) carboxylates form a 1D chain, which are further coordinated by pyridine groups from the neighboring ligands to result in the formation a 2D bilayer. These bilayers arising from the *syn* and *anti* bridging carboxylates are further packed in ···AA··· fashion in 3D space. The olefinic groups of the ligands are aligned parallel within the

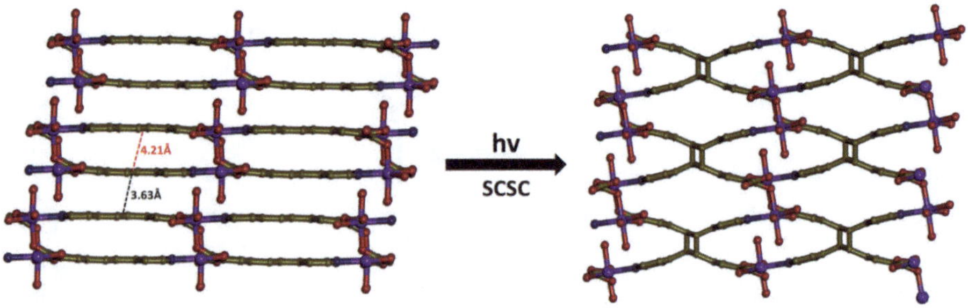

Fig. 23 SCSC photo-transformation of a 2D CP, **39**, to a 3D MOF

bilayer as well as between the bilayers in heat-to-tail manner. The distance between the olefinic groups within the bilayers is 4.21 Å, but in contrast, the distance between the olefinic groups between the layers is shorter (3.68 Å) as shown in Fig. 23. By irradiation under UV light, this 2D bilayer structure showed quantitative photo-cycloaddition between the layers in an SCSC manner and resulted in the formation of a 3D MOF. Interestingly, the coordinated water molecule does not destruct the single crystals during the cycloaddition reactions [50].

6.7 *Transformation of 2D Interdigitated CP to 3D MOF*

Terminal ligands have rarely been utilized to make photo-reactive CPs. In this example, the use of 4spy in linking the 2D layers to form interpenetrated 3D structures by [2+2] cycloaddition reaction is described [51]. Two isotypical interdigitated photo-reactive CPs, $[Zn_2(cca)_2(4spy)_2]$ (**40**) and $[Zn_2(ndc)_2(4spy)_2]$ (**41**), have been constructed from the paddle-wheel SBU, $\{Zn_2(O_2C–C)_4\}$. The linear dicarboxylate ligands 4-carboxycinnamate (cca) or 2,6-naphthalenedicarboxylate (ndc) generate the 2D layer structure with (4,4) connectivity. The two apical positions of each paddle-wheel SBU are coordinated by monodentate terminal 4spy ligands. Due to long arms of this ligand, the 2D layers are interdigitated with each other in the 3D space. The olefin groups of terminal 4spy ligands are aligned in heat-to-tail fashion between the first and fourth layers with a distance of 3.85 Å. Of these two 2D MOFs, only the cca derivate showed [2+2] cycloaddition in an SCSC fashion, and the final structure of photo reacted compound, $[Zn_2(cca)_2($*rctt*-4-ppcb$)]$ (**42**) (where *rctt*-4-ppcb = 1,3-bis(4′-pyridyl)-2,4-bis(phenyl)cyclobutane), has been determined to be a triply interpenetrated MOF with α-Po topology as shown in Fig. 24. But the single crystal was not reserved for 41 after the UV irradiation experiment. It is demonstrated that the terminal photo-reactive ligands can also be used for the structural transformation by [2+2] photo-cycloaddition reactions.

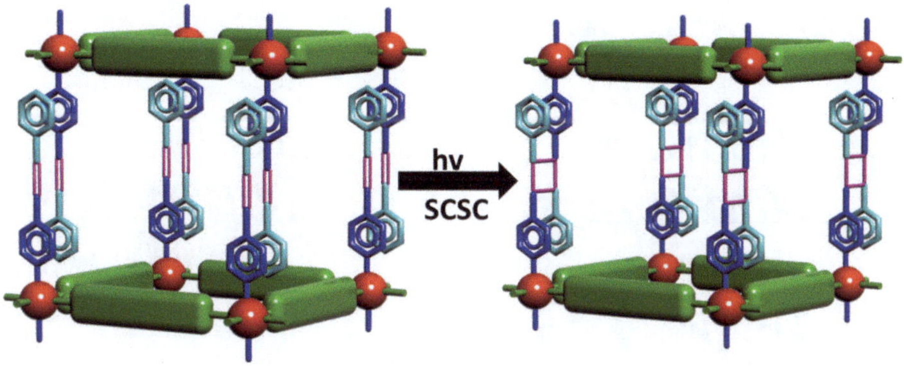

Fig. 24 Schematic representation of the transformation a 2D interdigitated CP to a triply interpenetrated 3D MOF. The interpenetration is not shown

6.8 Post-Synthetic Modification to Fine-Tune the Sorption Behavior of 2D Sheets

Kitagawa and coworkers have demonstrated how to control the reactivity and pore surface of MOFs by photo-irradiation [52]. They have synthesized an interdigitated 2D sheet $[Zn_2(abdc)_2(bpy)_2(DMF)_{1.5}]$ (**43**) using 5-amino-benzene-1,3-dicarboxylate (abdc) and 4,4′-bipyridine (bpy). The amino group is exposed on both sides of the pore which has been diazotized followed by azidation with NaN_3 to get azido derivative. Surprisingly, this MOF containing azide group is thermally stable up to 150 °C. But this is nonporous for oxygen sorption. On photo-irradiation at 77 K, the adsorbed amount of oxygen was significantly improved due to the loss of nitrogen and formation of triplet nitrene radicals in the pores. Their results created a strategy to produce PCPs with reactive sites such as radicals and carbenes which cannot be trapped by conventional synthetic conditions [52]. The details are depicted in Fig. 25.

7 Photo-Reactive 3D CPs

7.1 Post-Synthetic Modification of Double-Pillared MOFs by [2+2] Cycloaddition Reaction

Post-synthetic modification of double-pillared MOFs by [2+2] photo-cycloaddition was demonstrated in the following examples. Three photo-reactive double-pillared MOFs, $[Zn_2(bpe)_2(muco)_2]\cdot DMF\cdot H_2O$ (**44**), $[Zn_2(bpe)_2(bdc)_2]\cdot DMF$ (**45**), and $[Zn_2(bpe)_2(fum)_2]\cdot H_2O$ (**46**) (H_2muco = *trans,trans*-muconic acid, H_2bdc = 1,4-benzene dicarboxylic acid, and H_2fum = fumaric acid), have been investigated [53]. In these structures, the layer is formed by the $\{Zn_2(O_2C–C)_4\}$ SBU in which each

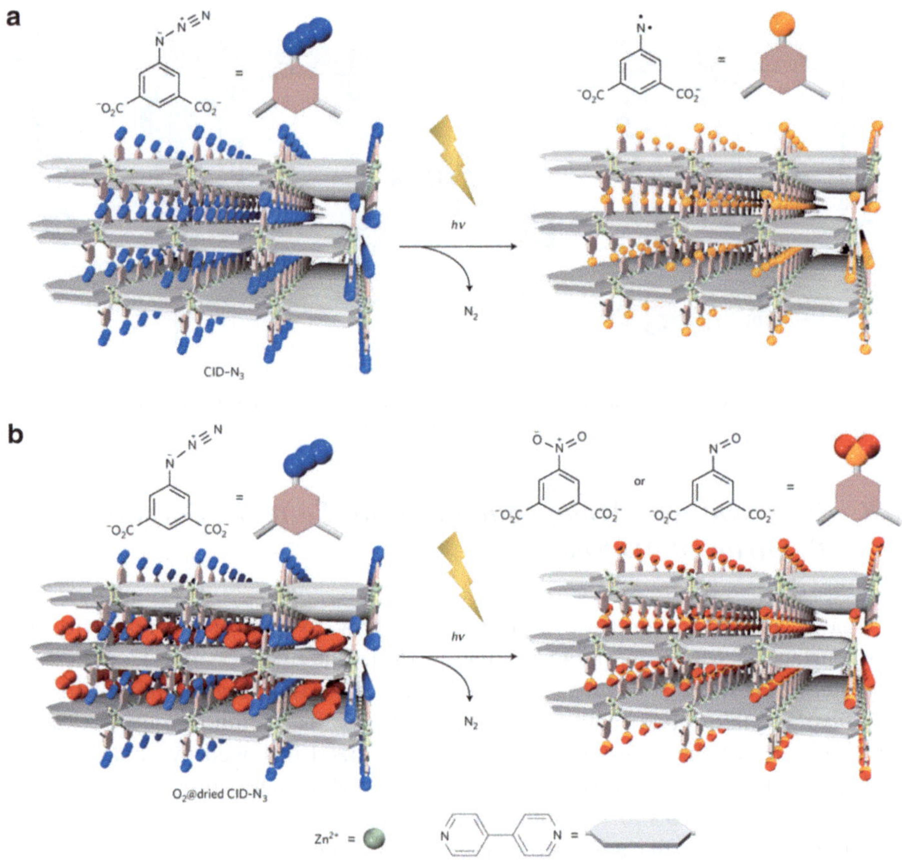

Fig. 25 Schematic diagram illustrating the photoactivation of 2D CP, **43**. (**a**) Photoactivation of the azide moieties leads to the formation of triplet nitrenes on the pore surface. (**b**) Photochemical trapping of physisorbed oxygen molecules in the framework of dried **43**. This figure is reproduced from [52] with permission

Zn(II) is chelated by a carboxylate and bridged by two carboxylates. These 2D layers are further double-pillared by the bpe ligands at the axial positions of Zn(II) atoms which are separated by 3.96 Å, to produce the 3D structure. This arrangement assists the bpe pairs to orient in parallel with the olefin groups separated by 3.79 Å (Fig. 26). Upon irradiation under UV light, all these MOFs showed quantitative cycloaddition reaction across the double pillars as confirmed by ^{1}H-NMR spectral studies. Among the three, MOFs containing muco and bdc ligands (**44** and **45**) underwent SCSC structural transformation. It has been shown that it is possible to modify the backbone of the spacer ligand without affecting the topology or porosity of the MOF during this structural transformation [53].

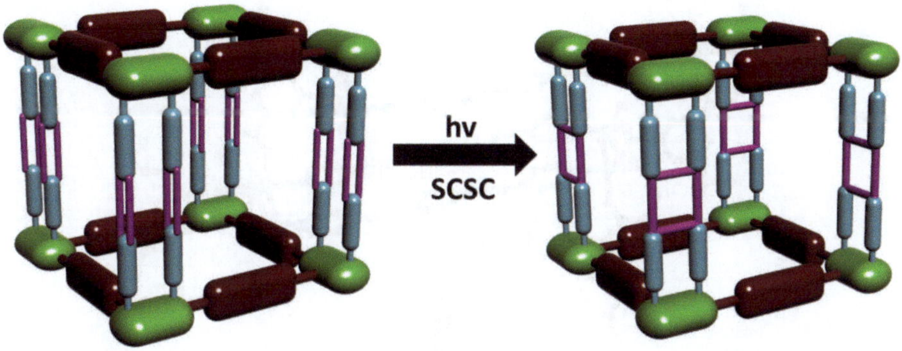

Fig. 26 Schematic representation of post-synthetic modification of 3D MOF using [2+2] photo-dimerization reaction

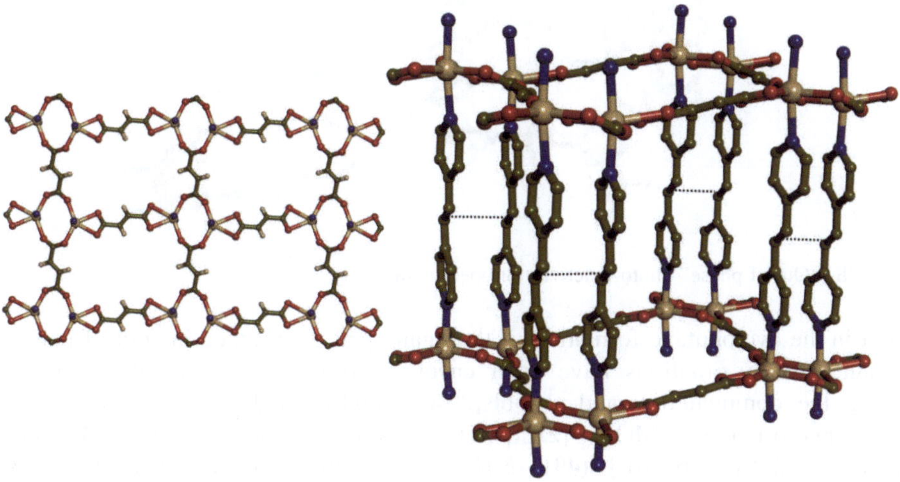

Fig. 27 A portion of the layer formed by Cd_2(fumarate)$_2$ (*left*) and the photo-reactive double-pillared MOF (**47**) (*right*)

A Cd(II) fumarate compound [Cd_2(bpe)$_2$(fum)$_2$] (**47**) similar to the Zn(II) analog with double-pillared MOF structure shown in Fig. 27 was also shown to be photo-reactive [54].

7.2 Photo-Reactive Double-Pillared MOFs Containing Conjugated Olefin Bonds

Conjugated C=C bonds containing spacer ligands are interesting due to many possibilities these ligands can offer in terms of photo-reactivity. When both C=C bonds are aligned "face-to-face" in parallel, the ligand can undergo double [2+2] cycloaddition reaction. One C=C bond from each ligand can also take

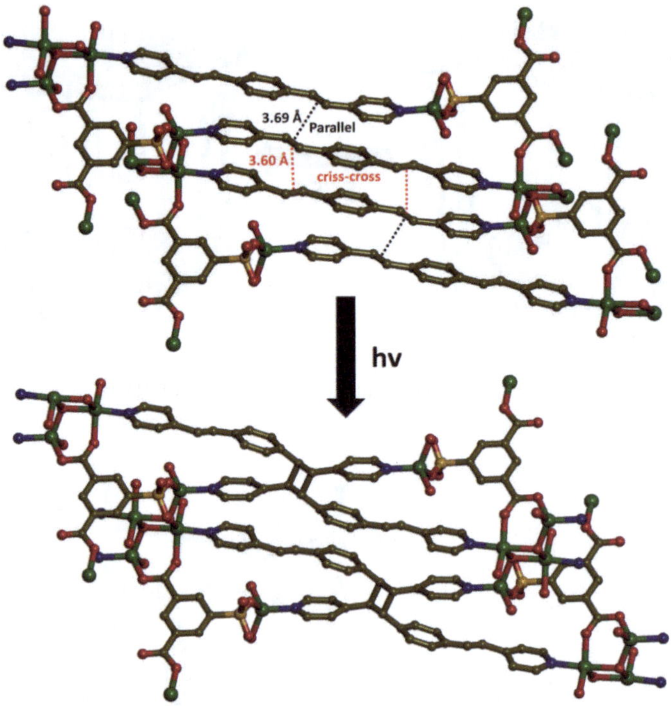

Fig. 28 "Out of phase" photo-reactivity of bpeb in **48**.

part in the cyclobutane formation, if the ligand pair is arranged in "out-of-phase" manner. Both situations have been encountered by Lang and coworkers by using the conjugated ligand, 1,4-bis[2-(4-pyridyl)ethenyl]benzene (bpeb) [55]. Two photo-reactive MOFs $[Zn_4(\mu_3\text{-}OH)_2(5\text{-sipa})_2(1,4\text{-bpeb})_2]\cdot 4H_2O$ (**48**) and $[Cd_2(1,3\text{-pda})_2(1,4\text{-bpeb})_2]$ (**49**) (5-H_3sipa = 5-sulfoisophthalic acid and 1,3-H_2pda = 1,3-phenylenediacetic acid) have been synthesized.

In the case of Zn(II) MOF, a 2D layer $[Zn_4(\mu_3\text{-}OH)_2(5\text{-sipa})_2]$ with (6,3) connectivity related CdI_2 is formed by linking six equivalent 5-sipa with three equivalent $\{Zn_4(\mu_3\text{-}OH)_2\}$. The axial positions of these $\{Zn_4(\mu_3\text{–}OH)_2\}$ sheets are further pillared by two pairs of bpeb ligands to form the double-pillared 3D network structure. However, each bpeb pillar is arranged in between two other bpeb ligands, where one C═C bond is aligned "in-phase" and the other in "out-of-phase," and all the olefin groups satisfy the topochemical distance criteria. However, the "in-phase" C═C bond pairs are crisscrossed and the "out-of-phase" double bonds have parallel orientation. As a result, UV irradiation of this compound showed quantitative photoreaction between the "out of phase" C═C pairs only and resulted in the formation of another 3D structure in an SCSC fashion (Fig. 28).

Similarly in the (4,4) sheets formed by the $\{Cd_2(1,3\text{-pda})_4\}$, the bpeb pairs bonded to the Cd(II) atoms are extended from either side of the sheets to link the neighboring $\{Cd_2(1,3\text{-pda})_4\}$ sheets to form another double-pillared 3D structure.

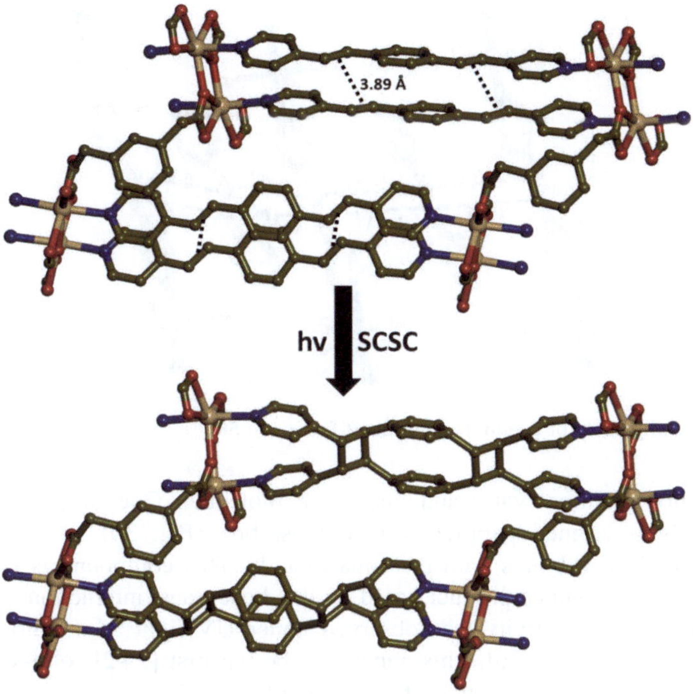

Fig. 29 "In-phase" photo-reactivity of bpeb in **49**

The bpeb ligand pairs are aligned in "in-phase" manner with the distances between C═C bonds, 3.886 and 4.023 Å. The structural transformation occurs in an SCSC manner upon UV irradiation of these crystals with the quantitative formation of double cyclobutane rings across bpeb pairs (Fig. 29). This is another elegant example for the post-synthetic modification of double-pillar ligands into a single pillar by [2+2] cycloaddition reaction [55].

7.3 Host–Guest Reactions in Pillared-Layer MOFs

An interesting alignment between the coordinated and guest bpe molecules in the channels that leads to photo-reactivity was described by Gao's laboratory [56]. A photo-reactive pillared-layer CP $[Mn_2(HCO_2)_3(bpe)_2(H_2O)_2]\ ClO_4{\cdot}H_2O{\cdot}bpe$ (**50**) has the layer structure formed by $Mn_2(HCO_2)_3$ linked by bpe pillars. Each Mn(II) is linked to three others through bridging formate ligands and an aqua ligand to form a layer, and the two axial sites are occupied by the bpe ligands. Two different types of channels formed in this 3D framework structure are occupied

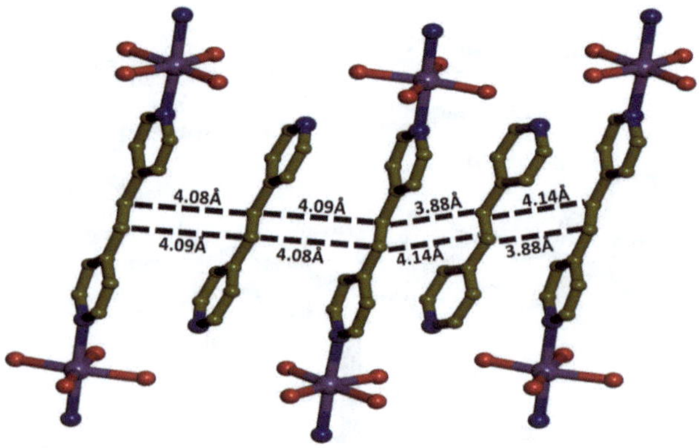

Fig. 30 The alignment of bpe in the host–guest fashion in **50**

by the anions, lattice water and guest bpe molecule. One of the coordinated bpe ligands is aligned parallel with a guest bpe (Fig. 30) stabilized by the O–H···N hydrogen bonds with the aqua ligands. This compound is expected to yield 66.7% cyclobutane product based on this host–guest interaction, and indeed about 60% photo-reactivity was observed under UV light as monitored by the ^{1}H-NMR spectroscopy [56]. This appears to be the first [2+2] solid-state photoreaction of a guest molecule with the framework.

Kitagawa and coworkers have recently described an interesting but different host–guest-polymerization concept to synthesize cross-linked polymers such as polystyrene, methylmethacrylate, and vinylacetate with pseudo-crystallinity in non-photochemical route [57]. In order to achieve this, they have first incorporated the cross-linker 2,5-divinyl-benzene-1,4-dicarboxylate (DVTP) into the porous CP [Cu(DVTP)(triethylenediamine)$_{0.5}$] (**51**). The host framework containing porous channels with dangling vinyl groups provides a suitable environment for radical polymerization of these monomers as shown in Fig. 31. This is obviously different from photopolymerization by [2+2] cycloaddition reaction.

7.4 Cis–Trans *Isomerization on the Gas Adsorption Property of MOFs*

Cis–trans isomerization of azobenzene has widely been exploited in photoresponsive materials [58, 59]. The pore structure of the MOF [Zn_2(bdc)$_2$(triethylenediamine)] (**52**) (pore size, 7.5 Å × 7.5 Å) has been shown to be deformed by the inclusion of certain aromatic guest molecules. This structural dynamic property of the MOF and the photo behavior of azobenzene have been nicely combined

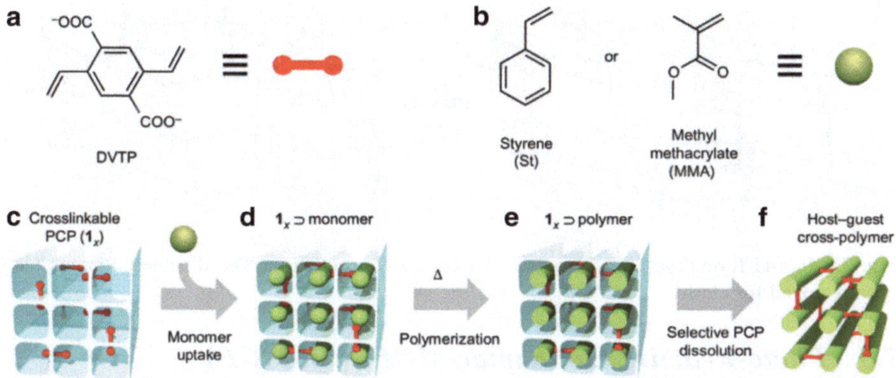

Fig. 31 Radical polymerization on the MOF surfaces. (**a**) Molecular structure of the DVTP ligand. (**b**) Monomers used in the host–guest cross-polymerizations. (**c**) Cross-linkable MOF [Cu(DVTP)$_x$(terephthalate)$_{1-x}$(triethylenediamine)$_{0.5}$] (**1**$_x$ shown in *light blue*) prepared in solid solution fashion. (**d**) Host **1**$_x$ accommodates vinyl monomers in its nanocavities. (**e**) Heat triggers radical cross-polymerization in the presence of AIBN initiator, yielding **1**$_x$ polymer nanocomposites. (**f**) Selective decomposition of the MOF matrix generates highly ordered cross-linked polymers. This figure is reproduced from [57] with permission

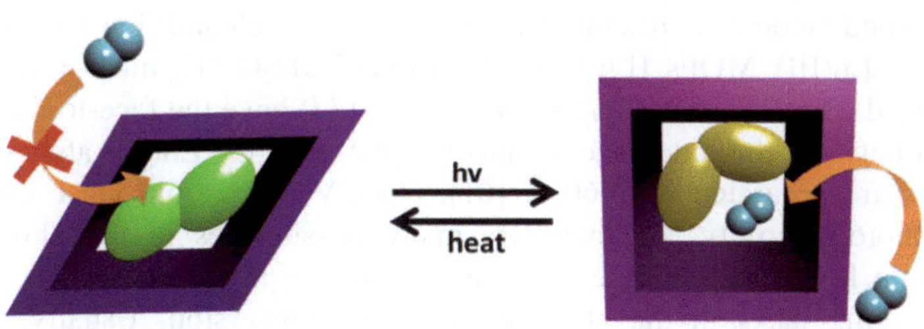

Fig. 32 Reversible transformation of **52** by photo isomerization of azobenzene

to study the *cis–trans* isomerization of azobenzene inside the pores of **52**. The change in the structure of MOF due to UV light triggered reversible *trans* to *cis* form and thermal isomerization of *cis* to *trans* form resulting in a drastic switching of the adsorption property of the MOF. The isomerization under UV light can easily be followed by the disappearance of the frequency at 690 cm^{-1} due to the *trans* isomer and appearance of a new frequency at 697 cm^{-1} due to the formation of *cis* isomer in the IR spectroscopy. This isomerization is accompanied by the changes of the framework structure from orthorhombic to the tetragonal form (Fig. 32). This dramatic change in the structure also influences the nitrogen adsorption behavior [60].

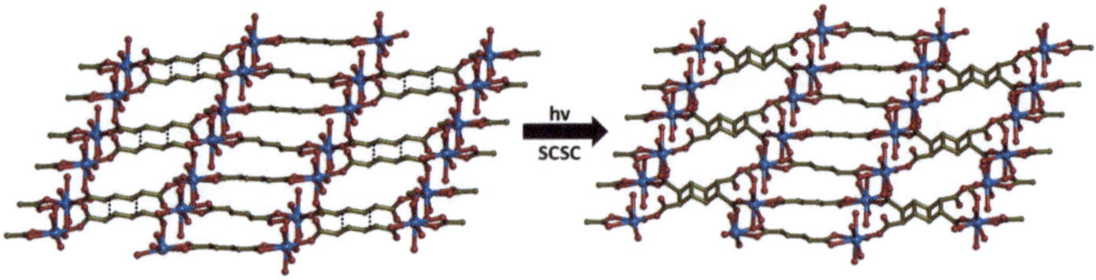

Fig. 33 Partial formation of ladderane in **53** upon UV irradiation. The disordered atoms have been removed for clarity

7.5 Photo-Reactive Lanthanide-MOFs (Ln-MOF)

Lanthanum(III) atoms have rarely been used to study the photo-reactivity in CPs. Michaelides and coworkers explored the photo-reactivity of muconate ligands bonded to Er(III) and Y(III) atoms. Muconate is another linear spacer ligand with two conjugated C=C bonds in the backbone which presents different possibilities of photoproducts, including mono [2+2] cycloaddition or [4+4] cycloaddition products. The transformation of cyclobutane to cyclooctadiene by Cope rearrangement is another possibility. Further, it may also result in the formation of a highly strained ladderane structure by double [2+2] cycloaddition reaction.

The two Ln(III) MOFs $[Ln_2(muco)_3(H_2O)_6]\cdot 2.5H_2O$ (H_2muco = *trans,trans*-muconic acid), Ln(III) = Er(III) (**53**) and Y(III) (**54**) have the face-to-face parallel alignment between the muconate ligands templated by the Ln(III) atoms in which the olefins are separated by 3.60 Å [61]. On UV irradiation, these compounds showed photo-reactivity. As there are many possibilities in the photoreaction of muconate ligand as discussed before, identifying the photoproduct is one of the challenging tasks in the absence of SCSC conversion. Usually, ^{1}H-NMR spectroscopy has been used to monitor the nature of products in solution. Due to the insoluble nature of these products, they are normally digested with acids to solubilize the organic moiety, and there is a possibility for these products to rearrange from one structure to another under these conditions in solution. Interestingly, the photo-irradiation of Er(III) showed formation of ladderane structure with double [2+2] photo-cycloaddition reaction as shown in Fig. 33, which has been confirmed by partial SCSC transformation, where the single crystals are stable up to 55% of photoreaction. This highly strained ladderane formation has also been supported by other experimental evidences. This may be due to the generation of the stress created during the photoreaction, the single crystals were collapsed after 55% photoreaction [61]. Similarly, isomorphs of the above structures have also been investigated by the same authors [62]. In all these compounds, the separation distance between the olefin bonds varies from 3.7 to 4.4 Å and also not aligned in parallel. However, all these compounds participated in the photoreaction as confirmed by ^{1}H-NMR spectral studies.

8 [2+2] Cycloaddition Reaction to Monitor the Structural Transformations in CPs

8.1 Anisotropic Movements of 1D CPs by Desolvation

Since close proximity is essential for the [2+2] cycloaddition reaction, only ladder structure in which the C=C bond incorporated spacer ligands occupy the poles can be photo-reactive among the 1D CPs. Hence, the existence or the formation of ladder structures with these photo-reactive ligands can easily be monitored by the ^{1}H-NMR spectroscopy for the formation of cyclobutane rings after irradiating them under UV light, if the single crystal data are not available. The photo-reactivity of the ladder CPs has been used as a tool to understand the anisotropic molecular movements of CPs [43]. A 1D CP, [Ag(μ-bpe)-(H_2O)](TFA)·CH_3CN (**55**), has been trapped as a kinetic product during the synthesis of 1D ladder CP, [Ag_2(bpe)$_2$(TFA)$_2$] (**56**). In the linear CP, the Ag(I) has a T-shaped geometry from two bpe spacers and an aqua ligand. The hydrogen atoms of the coordinated water molecule are H-bonded to the O atoms of noncoordinated TFA anions and form a 2D H-bonded sheet-like structure. The olefin groups of the bpe are far away (5.15 Å) to satisfy the topochemical distance criterion. Surprisingly, the aqua ligand can easily be lost in air along with CH_3CN, and the single crystals were destroyed very easily. However, UV irradiation of this desolvated powder showed quantitative formation of cyclobutane rings as confirmed by ^{1}H-NMR data. The loss of coordinatively bonded aqua ligands at such a low temperature suggests some kinds of structural transformation taken place in the solid state. This photo-reaction confirms the formation of a ladder-like structure from the linear chains. All these suggest that the linear chain is a kinetic product and cooperative anisotropic molecular movements of the adjacent 1D chains occur during the desolvation process leading to the formation of the thermodynamically stable ladder CP (Fig. 34), which is stabilized by Ag···Ag and face-to-face π···π interactions.

8.2 Transformation of a Linear CP to a Ladder Structure by Thermal Dehydration

Another structural rearrangement of a linear CP to a ladder structure was also successfully monitored by the [2+2] cycloaddition reactions [44]. The Cd(II) center in the linear CP [Cd(bpe)(CH_3COO)$_2$(H_2O)] (**57**) has hepta-coordinated pentagonal bipyramidal geometry coordinated to two bpe ligands in *trans* fashion and chelated by two acetate groups and a water molecule. These linear chains are arranged in parallel one below another to form a plane with the help of H-bonds between the coordinated water molecule and chelated acetate O atoms from the

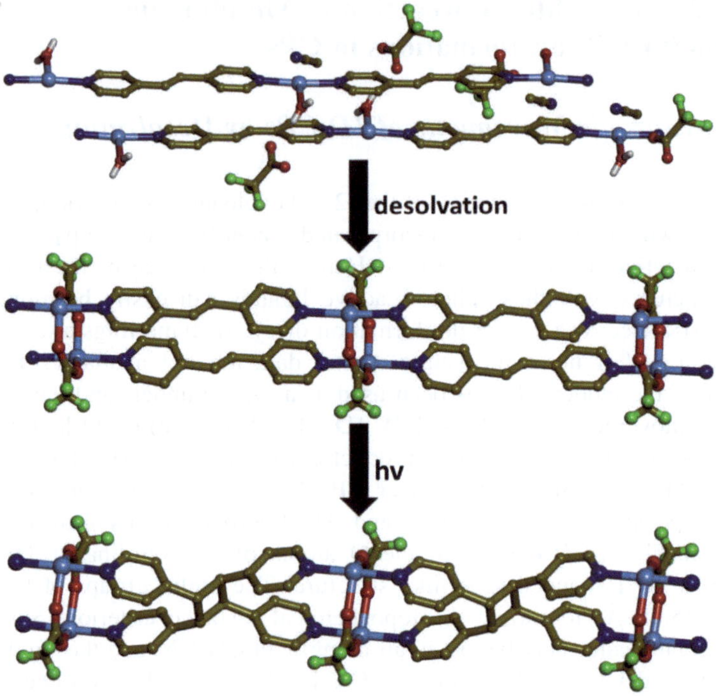

Fig. 34 Rearrangement of a linear CP to a photo-reactive ladder structure by desolvation

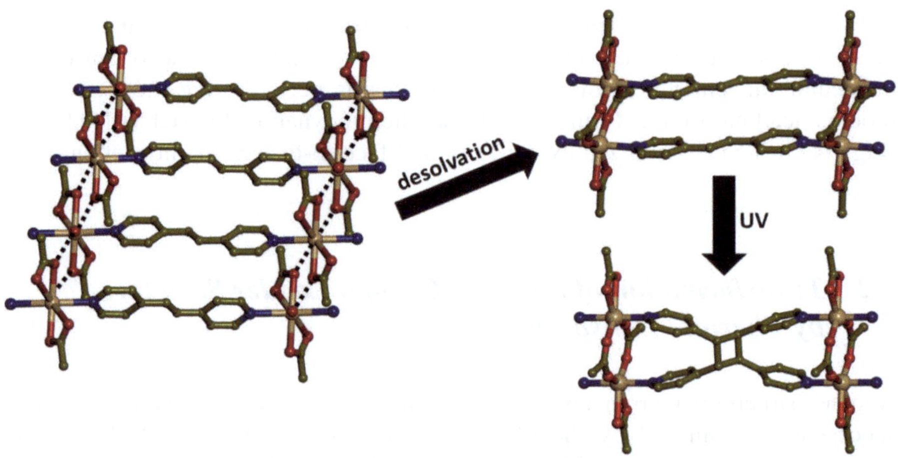

Fig. 35 Structural transformation of **57** upon desolvation and its photo-reactivity upon UV irradiation

adjacent chains as shown in Fig. 35. A close examination of the packing reveals that the closest bpe pairs are crisscrossed and the distance between these olefinic groups is 4.33 Å and hence expected to be a photo stable solid [63]. However, this compound showed partial photo-cycloaddition of bpe (33%) when exposed to UV irradiation for 30 h, and longer irradiation did not increase the formation of cyclobutane rings as monitored by ^{1}H-NMR spectroscopy. Interestingly, the TGA showed that the loss of coordinated water occurs in the range, 67–106°C. The loss of coordinated water can bring the adjacent chains closer to form a new bond between Cd(II) and oxygen atom of the adjacent chains, which are separated by 4.44 Å. During this structural transformation, formation of a stable structure might have forced the loss of coordinated water below its boiling point as observed by TGA. The dehydrated product exhibited 100% photo-reactivity when exposed to UV light 30 h. This photo-cycloaddition strongly supports the migration of acetates after the loss of coordinated solvent, in bringing the adjacent chains closer to form a ladder structure over 2D sheets. Therefore, the formation of cyclobutane rings can be used as a tool to follow the alignment of the double bonds during structural transformation.

8.3 Loss of Water Chains Transforms a Linear CP to a Ladder Structure

During the synthesis of $[Pb_2(\mu\text{-bpe})_2(\mu\text{-}O_2C\text{–}C_6H_5)_2(O_2C\text{–}C_6H_5)_2]$ (**13**) (see Sect. 3.1), a linear CP $[Pb(\mu\text{-bpe})(O_2C\text{–}C_6H_5)_2]\cdot 2H_2O$ (**58**) was trapped as a kinetic product [34]. In the solid-state structure, the guest water molecules aggregate to form a hydrogen-bonded 1D polymer. When this water aggregate was removed by thermal dehydration, it transforms to the photo-reactive ladder structure as confirmed by PXRD and UV irradiation experiments (Fig. 36). This work illustrates the generality of the design principles of synthesizing double-stranded ladder-like 1D CPs to align infinite C=C bond pairs for solid-state photo-reactivity by clipping *hemidirected* Pb(II) centers in a serial fashion.

9 Photopolymerization of Alkenoate Systems in CPs

Solid-state polymerization of Li(I), Na(I), and K(I) acrylates has been studied for many years, and such reactions have been induced by ^{60}Co γ-rays. Later, the polymerization of Ca(II), Ba(II), and Zn(II) methacrylates, which are of 1D and 2D CPs, was also reported [64–72]. In addition, some of these compounds also undergo polymerization by heating. However, the final structures of these compounds are not known since they do not retain their single crystallinity after the polymerization reaction.

10 Photo-Reactivity in Solution

A number of reactions conducted in solution including hydrothermal/solvothermal process produced CPs and MOFs containing photoproducts as evidence for the photo-reactivity in solution. It is clear that the metal ions, anions, and the co-ligands

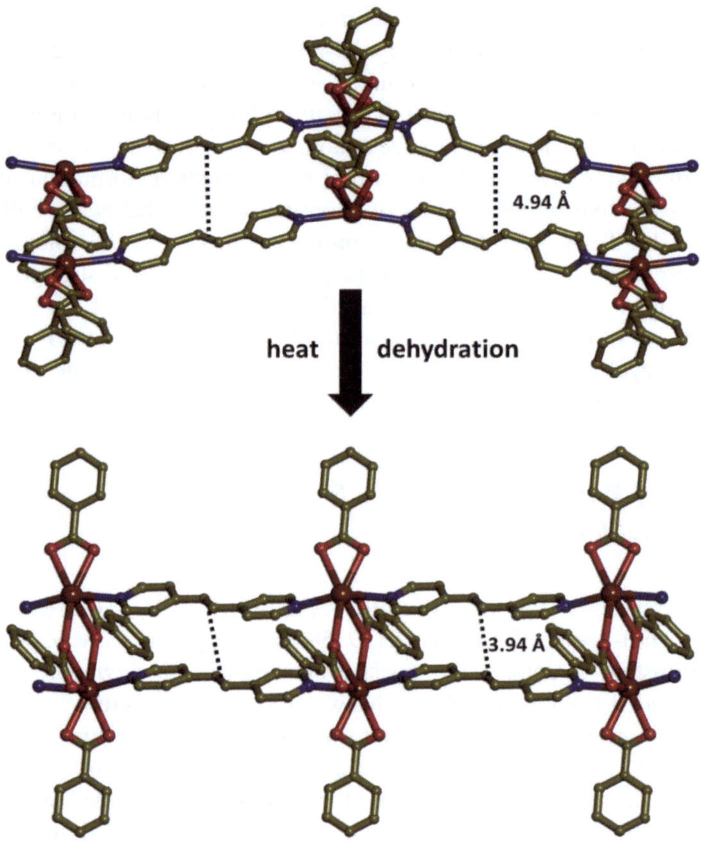

Fig. 36 Transformation of a linear CP, **17**, to a photo-reactive ladder structure, **13**, upon dehydration

assist in the formation of the stereoselective formation of cyclobutane derivatives in the novel solid-state structures. Once formed, the organic photoproduct can combine with the metal ions, anions, and co-ligands to form interesting CPs and MOFs [49–52]. But this area of research is out of scope of this chapter and hence not discussed further.

11 Conclusion

In this chapter, various photochemical reactions involving CPs and MOFs have been described. These include structural transformations, post-synthetic modification of surfaces to create reactive sites, radicals which are not possible by convention synthetic routes, polymerization on the surfaces, and *cis–trans* isomerization of the guest molecules in the cavities using the dynamic behavior of the structures. Further, [2+2] cycloaddition reaction has been used as a tool to monitor the changes in the solid-state structures in the absence of crystal structures. A number of polymerization reactions such as alkenoates have been investigated. The structures of the resultant products were hampered by the lack of suitable structural tool other than single crystal X-ray crystallography. On the other hand, most of the examples discussed in this chapter do maintain the single crystals at the end of the reactions making them amenable for structural elucidation to understand the reactivity completely.

Acknowledgements We thank the National University of Singapore for their continuous support. The Ministry of Education, Singapore, is gratefully thanked for financial support through NUS FRC Grant R-143-000-439-112. We also thank the Nature Publications for providing the copyright permissions for Figs. 25 and 31.

References

1. Ramamurthy V, Venkatesan K (1987) Photochemical reactions of organic crystals. Chem Rev 87(2):433–481
2. Theocharis CR, Jones W (1987) Organic solid state chemistry. In: Desiraju GR (ed) Organic solid state chemistry. Elsevier, Amsterdam, pp 47–68
3. Green BS, Lahav M, Rabinovich D (1979) Asymmetric synthesis via reactions in chiral crystals. Acc Chem Res 12(6):191–197
4. Jones W (ed) (1997) Organic molecular solids: properties and applications. CRC, Boca Raton
5. Toda F (ed) (2005) Organic solid state reactions. Topics in current chemistry, vol 254. Springer, Berlin
6. Keating AE, Garcia-Garibay MA (1998) Organic and inorganic photochemistry. In: Ramamurthy V, Schanze KS (eds) Organic and inorganic photochemistry. Marcel Dekker, New York, pp 195–248
7. Tanaka K, Toda F (2000) Solvent-free organic synthesis. Chem Rev 100(3):1025–1074
8. Toda F (1995) Solid state organic chemistry: Efficient reactions, remarkable yields, and stereoselectivity. Acc Chem Res 28(12):480–486
9. Lee-Ruff E, Mladenova G (2003) Enantiomerically pure cyclobutane derivatives and their use in organic synthesis. Chem Rev 103(4):1449–1484
10. Namyslo JC, Kaufmann DE (2003) The application of cyclobutane derivatives in organic synthesis. Chem Rev 103(4):1485–1538
11. Nishimura J, Nakamura Y, Hayashida Y, Kudo T (2000) Stereocontrol in cyclophane synthesis: a photochemical method to overlap aromatic rings. Acc Chem Res 33(10):679–686
12. Cohen MD, Schmidt GMJ, Sonntag FI (1964) Topochemistry. Part II. The photochemistry of trans-cinnamic acids. J Chem Soc 384:2000–2013
13. Schimdt GMJ (1971) Photodimerization in the solid state. Pure Appl Chem 27:647–678
14. Desiraju GR, Vittal JJ, Ramanan A (eds) (2011) Crystal engineering. World Scientific, Singapore
15. Kole GK, Vittal JJ (2013) Solid-state reactivity and structural transformations involving coordination polymers. Chem Soc Rev 42(4):1755–1775
16. MacGillivray LR, Papaefstathiou GS, Friščić T, Hamilton TD, Bučar D-K, Chu Q, Varshney DB, Georgiev IG (2008) Supramolecular control of reactivity in the solid state: from templates to ladderanes to metal – organic frameworks. Acc Chem Res 41(2):280–291
17. Biradha K, Santra R (2013) Crystal engineering of topochemical solid state reactions. Chem Soc Rev 42(3):950–967

18. Nagarathinam M, Peedikakkal AMP, Vittal JJ (2008) Stacking of double bonds for photochemical [2+2] cycloaddition reactions in the solid state. Chem Commun 42:5277–5288
19. Nagarathinam M, Vittal JJ (2006) A rational approach to crosslinking of coordination polymers using the photochemical [2+2] cycloaddition reaction. Macromol Rapid Commun 27(14):1091–1099
20. Vittal JJ (2007) Supramolecular structural transformations involving coordination polymers in the solid state. Coord Chem Rev 251(13–14):1781–1795 (special issue)
21. Praetorius P (1909) Thesis, Halle
22. Praetorius P, Korn F (1910) Ber Dtsch Chem Ges 43:2744
23. Stobbe H, Färber E (1925) Ber Dtsch Chem Ges 58:1548
24. Alcock NW, de Meester P, Kemp TJ (1979) Solid-state photochemistry. Part 1. Nature of the stereocontrol in the photodimerisation of dibenzylideneacetone by UO22+ ion: Crystal and molecular structure of trans-dichlorobis(trans, trans-dibenzylideneacetone)dioxouranium(VI) and of its acetic acid solvate. J Chem Soc Perkin Trans 2(7):921–926
25. Theocharis CR (1987) Co-ordination polymers based on 2,5-dibenzylidenecyclopentanone, which are photochemically cross-linkable. J Chem Soc Chem Commun 0(2):80–81
26. Chu Q, Swenson DC, MacGillivray LR (2005) A single-crystal-to-single-crystal transformation mediated by argentophilic forces converts a finite metal complex into an infinite coordination network. Angew Chem Int Ed 44(23):3569–3572
27. Harada J, Ogawa K (2001) Invisible but common motion in organic crystals: A pedal motion in stilbenes and azobenzenes. J Am Chem Soc 123(44):10884–10888
28. Dutta S, Bucar D-K, Elacqua E, MacGillivray LR (2013) Single-crystal-to-single-crystal direct cross-linking and photopolymerisation of a discrete Ag(I) complex to give a 1D polycyclobutane coordination polymer. Chem Commun 49(11):1064–1066
29. Peedikakkal AMP, Koh LL, Vittal JJ (2008) Photodimerization of a 1D hydrogen-bonded zwitter-ionic lead(ii) complex and its isomerization in solution. Chem Commun 4:441–443
30. Paul AK, Karthik R, Natarajan S (2011) Synthesis, structure, photochemical [2+2] cycloaddition, transformation, and photocatalytic studies in a family of inorganic–organic hybrid cadmium thiosulfate compounds. Cryst Growth Des 11(12):5741–5749
31. Peedikakkal AMP, Vittal JJ (2008) Solid-state photochemical [2+2] cycloaddition in a hydrogen-bonded metal complex containing several parallel and crisscross C=C bonds. Chem Eur J 14(17):5329–5334
32. Toh NL, Nagarathinam M, Vittal JJ (2005) Topochemical photodimerization in the coordination polymer [{(CF3CO2)(μ-O2CCH3)Zn}2(μ-bpe)2]n through single-crystal to single-crystal transformation. Angew Chem Int Ed 44(15):2237–2241
33. Nagarathinam M, Vittal JJ (2010) Solid-state synthesis of coordination polymers for [2+2] photoreactions by grinding. Aust J Chem 63(4):589–595
34. Kole GK, Peedikakkal AMP, Toh BMF, Vittal JJ (2013) Solid-state structural transformations and photoreactivity of 1D-ladder coordination polymers of PbII. Chem Eur J 19(12): 3962–3968
35. Liu D, Li N-Y, Lang J-P (2011) Single-crystal to single-crystal transformation of 1D coordination polymer via photochemical [2+2] cycloaddition reaction. Dalton Trans 40(10): 2170–2172
36. Papaefstathiou GS, Zhong Z, Geng L, MacGillivray LR (2004) Coordination-driven self-assembly directs a single-crystal-to-single-crystal transformation that exhibits photocontrolled fluorescence. J Am Chem Soc 126(30):9158–9159
37. Papaefstathiou GS, Georgiev IG, Friscic T, MacGillivray LR (2005) Directed assembly and reactivity of olefins within a one-dimensional ladder-like coordination polymer based on a dinuclear Zn(ii) platform. Chem Commun 31:3974–3976
38. Ou Y-C, Liu W-T, Li J-Y, Zhang G-G, Wang J, Tong M-L (2011) Solvochromic and photodimerization behaviour of 1D coordination polymer via single-crystal-to-single-crystal transformation. Chem Commun 47(33):9384–9386

39. Santra R, Banerjee K, Biradha K (2011) Weak AgAg and Ag[small pi] interactions in templating regioselective single and double [2+2] reactions of N, N[prime or minute]-bis (3-(4-pyridyl)acryloyl)-hydrazine: Synthesis of an unprecedented tricyclohexadecane ring system. Chem Commun 47(38):10740–10742
40. Eubank JF, Kravtsov VC, Eddaoudi M (2007) Synthesis of organic photodimeric cage molecules based on cycloaddition via metal–ligand directed assembly. J Am Chem Soc 129(18):5820–5821
41. Liu D, Lang J-P, Abrahams BF (2013) Stepwise ligand transformations through [2+2] photodimerization and hydrothermal in situ oxidation reactions. Chem Commun 49(26): 2682–2684
42. Peedikakkal AMP, Vittal JJ (2010) Solid-state photochemical behavior of a triple-stranded ladder coordination polymer. Inorg Chem 49(1):10–12
43. Nagarathinam M, Vittal JJ (2006) Anisotropic movements of coordination polymers upon desolvation: solid-state transformation of a linear 1D coordination polymer to a ladderlike structure. Angew Chem Int Ed 45(26):4337–4341
44. Nagarathinam M, Vittal JJ (2008) Photochemical [2+2] cycloaddition as a tool to study a solid-state structural transformation. Chem Commun 4:438–440
45. Ou YC, Zhi DS, Liu WT, Ni ZP, Tong ML (2012) Single-crystal-to-single-crystal transformation from 1D staggered-sculls chains to 3D NbO-type metal-organic framework through [2+2] photodimerization. Chem Eur J 18(24):7357–7361
46. Michaelides A, Skoulika S, Siskos MG (2004) Assembly of a photoreactive coordination polymer containing rectangular grids. Chem Commun 21:2418–2419
47. Miao X-H, Zhu L-G (2010) Regiocontrolled [2+2] photodimerization of Cd(II) metal complexes in both solution and solid state. Dalton Trans 39(6):1457–1459
48. Chanthapally A, Kole GK, Qian K, Tan GK, Gao S, Vittal JJ (2012) Thermal cleavage of cyclobutane rings in photodimerized coordination-polymeric sheets. Chem Eur J 18(25): 7869–7877
49. Sato H, Matsuda R, Mir MH, Kitagawa S (2012) Photochemical cycloaddition on the pore surface of a porous coordination polymer impacts the sorption behavior. Chem Commun 48(64):7919–7921
50. Xie M-H, Yang X-L, Wu C-D (2011) From 2D to 3D: a single-crystal-to-single-crystal photochemical framework transformation and phenylmethanol oxidation catalytic activity. Chem Eur J 17(41):11424–11427
51. Medishetty R, Koh LL, Kole GK, Vittal JJ (2011) Solid-state structural transformations from 2D interdigitated layers to 3D interpenetrated structures. Angew Chem Int Ed 50(46): 10949–10952
52. Sato H, Matsuda R, Sugimoto K, Takata M, Kitagawa S (2010) Photoactivation of a nanoporous crystal for on-demand guest trapping and conversion. Nat Mater 9(8):661–666
53. Mir MH, Koh LL, Tan GK, Vittal JJ (2010) Single-crystal to single-crystal photochemical structural transformations of interpenetrated 3D coordination polymers by [2+2] cycloaddition Reactions13. Angew Chem Int Ed 49(2):390–393
54. Michaelides A, Skoulika S, Siskos MG (2008) Designed self-assembly of a reactive metal-organic framework with quasi [small alpha]-Po topology. CrystEngComm 10(7):817–820
55. Liu D, Ren Z-G, Li H-X, Lang J-P, Li N-Y, Abrahams BF (2010) Single-crystal-to-single-crystal transformations of two three-dimensional coordination polymers through regioselective [2+2] photodimerization reactions. Angew Chem Int Ed 49(28):4767–4770
56. Wang X-Y, Wang Z-M, Gao S (2007) A pillared layer MOF with anion-tunable magnetic properties and photochemical [2+2] cycloaddition. Chem Commun 11:1127–1129
57. Distefano G, Suzuki H, Tsujimoto M, Isoda S, Bracco S, Comotti A, Sozzani P, Uemura T, Kitagawa S (2013) Highly ordered alignment of a vinyl polymer by host–guest cross-polymerization. Nat Chem 5(4):335–341
58. Beharry AA, Woolley GA (2011) Azobenzene photoswitches for biomolecules. Chem Soc Rev 40(8):4422–4437

59. Kumar GS, Neckers DC (1989) Photochemistry of azobenzene-containing polymers. Chem Rev 89(8):1915–1925
60. Yanai N, Uemura T, Inoue M, Matsuda R, Fukushima T, Tsujimoto M, Isoda S, Kitagawa S (2012) Guest-to-host transmission of structural changes for stimuli-responsive adsorption property. J Am Chem Soc 134(10):4501–4504
61. Michaelides A, Skoulika S, Siskos MG (2011) Photoreactive 3D microporous lanthanide MOFs: formation of a strained ladderane in a partial single crystal-to-single crystal manner. Chem Commun 47(25):7140–7142
62. Michaelides A, Skoulika S, Siskos MG (2013) 2D And 3D photoreactive lanthanide MOFs of trans, trans-muconic acid. Chem Commun 49(10):1008–1010
63. Schmidt GMJ (1971) Pure Appl Chem 27:647–678
64. Vela MJ, Snider BB, Foxman BM (1998) Solid-state polymerization of aquabis(3-butenoato) calcium. Chem Mater 10(10):3167–3171
65. Inoki M, Akutsu F, Kitayama Y, Kasashima Y, Naruchi K (1998) Thermal polymerizations of alkali and alkaline earth 4-vinylbenzoates in bulk. Macromol Chem Phys 199(4):619–623
66. Kudoh M, Akutsu F, Nakaishi E, Kobayashi T, Naruchi K, Miura M (1994) Thermal polymerization of alkali and zinc methacrylates in the solid state. Macromol Rapid Commun 15(3): 239–242
67. Bowden MJ, O'Donnell JH, Sothman RD (1969) Radiation induced solid state polymerization of derivatives of methacrylic acid. VI. Polymerization of barium methacrylate dihydrate during irradiation. Macromol Chem Phys 122(1):186–195
68. O'Donnell JH, Sothman RD (1968) Radiation-induced, solid-state polymerization of derivatives of methacrylic acid. I. Postirradiation polymerization of zinc methacrylate. J Polym Sci Part A Polym Chem 6(5):1073–1086
69. Costaschuk FM, Gilson DFR, St. Pierre LE (1971) The solid-state polymerization of hydrated barium methacrylate. Macromolecules 4(1):16–19
70. Costaschuk FM, Gilson DFR, St. Pierre LE (1970) Solid-state polymerization of hydrated calcium acrylate. Macromolecules 3(4):393–397
71. Alcock NW, de Meester P, Kemp TJ (1979) Solid-state photochemistry. Part 1. Nature of the stereocontrol in the photodimerisation of dibenzylideneacetone by UO22+ ion: Crystal and molecular structure of trans-dichlorobis(trans, trans-dibenzylideneacetone)dioxouranium (VI) and of its acetic acid solvate. J Chem Soc Perkin Trans 2 0(7):921–926
72. Vela MJ, Buchholz V, Enkelmann V, Snider BB, Foxman BM (2000) Solid-state polymerization of bis(but-3-enoato)zinc: The generation of a stereoregular oligomer. Chem Commun 0(22):2225–2226

Struct Bond (2014) 157: 145–166
DOI: 10.1007/430_2013_107

Published online: 29 May 2013

Metal–Organic Frameworks for Second-Order Nonlinear Optics

Shaowu Du and Huabin Zhang

Abstract In this chapter, we highlight recent advances and perspectives in the design and synthesis of NLO-active metal–organic frameworks (MOFs). Some of these compounds show impressive second harmonic generation (SHG) responses and may have potential use in NLO-applications. Noncentrosymmetric MOFs can be synthesized mainly by using chiral ligands, unsymmetrical achiral ligands, and mixed metal ions. Organic ligands containing donor-π-acceptor systems can be incorporated into MOFs to improve the SHG activity of the bulk materials through push–pull effect. Some diamondoid or octupolar MOFs have been shown to exhibit excellent SHG properties. Besides, a number of MOFs with 2D grid-type or 1D helical chain structures display good SHG activity.

Keywords Diamondoid net · Metal–organic framework · Noncentrosymmetric MOFs · Nonlinear optics · Octupolar symmetry · Push–pull effect · Second harmonic generation

Contents

S. Du (✉) and H. Zhang
Fujian Institute of Research on the Structure of Matter, Chinese Academy of Sciences, Fujian, Fuzhou 350002, P. R. China
e-mail: swdu@fjirsm.ac.cn

Abbreviations

1,3-bimb	1,3-Bis(imidazol-1-ylmethyl)-benzene
1,4-bimb	1,4-Bis(imidazol-1-ylmethyl)-benzene
4-pya	4-Pyridylacrylate
aptz	5-(6-Aminopyridin-3-yl)tetrazol-1-ide
BCDC	*N,N′*-bis(4-cyanophenyl)-(1*R*,2*R*)-diaminocyclohexane
cda	Carbamyldicyanomethanide anion
D-H_2ca	D-camphoric acid
DMF	*N,N*-dimethylformamide
DPASD	4-(4-(Diphenylamino)styryl)-1-dodecylpyridinium
dpys	4,4′-Dipyridylsulfide
DSAT	4-*N,N*-dimethylamino-4′-*N*′-methyl-stilbazolium tosylate
(*E*)-4-pyv-3-bza	(*E*)-3-(2-(4-Pyridyl)vinyl)benzoate
(*E*)-4-pyv-4-bza	(*E*)-4-(2-(4-pyridyl)vinyl)benzoate
H_2dnty	3,5-Dinitrotyrosine
H_2nicO	6-Hydroxynicotinic acid
H_2oba	4,4′-Oxybis(benzoic acid)
H_2SCMC	*S*-carboxymethyl-L-cysteine
H_2sdba	4,4′-Sulfonyldibenzoic acid
H_3tzpbin	(1*R*,2*R*)-1-(3-(1*H*-tetrazol-5-yl)phenyl)-2-(pyridin-4-yl)-2,3-dihydro-1*H*-benzo[*e*]indole
HAmidn	2-Amino-4,5-imidazoledicarbonitrile
H-Imazethapyr	(2-(4,5-Dihydro-4-methyl-4-(1-methylethyl)-5-oxo-1*H*-imidazol-2-yl)-5-ethyl-3-pyridinecarboxylic acid
Htrtr	3-(1,2,4-Triazole-4-yl)-1*H*-1,2,4-triazole
Htzpbin	1-(3-(1*H*-tetrazol-5-yl)phenyl)-2-(pyridin-4-yl)-3*H*-benzo[*e*]indole
inic	Isonicotinate
m-H_2bdc	1,3-Benzenedicarboxylic acid
nic	Nicotinate
OH-*m*-H_2bdc	5-Hydroxyisophthalic acid
phen	1,10-Phenanthroline
pyb	4-(4-Pyridyl)benzoate
quin-6-c	Quinoline-6-carboxylate
spcp	4-Sulfanylmethyl-4′-phenylcarboxylate pyridine
THF	Tetrahydrofuran

1 Introduction

The development of NLO materials can be traced back to the first observation of second-harmonic generation (SHG) by Franken et al. in 1961 in a quartz crystal when a ruby laser was used as incident light resource [1]. Since then, intensive research has been carried out in the field of NLO materials as they can expand the range of laser wavelengths. At present, NLO have been identified as excellent candidates for emerging photonic technologies based on the fact that photons are capable of processing information with the speed of light. They also possess vast potential for use in a variety of photonic systems, including high-speed optical modulators, ultra-fast optical switches, and high-density optical data storage media. These devices are essential for continued advancement in the effort to transform information storage and transmission from electrical to the optical regime [2–4].

In the beginning, most interest in NLO mateirals was focused on inorganic materials, such as quartz, potassium dihydrogen phosphate (KDP), lithium niobate ($LiNbO_3$), potassium titanyl phosphate (KTP), β-barium borate and semiconductors such as cadmium sulfide, selenium, and tellurium. Although many of them have been widely used in commercial frequency conversion for lasers and optical parametric generator, the choice of these materials is still rather limited [5, 6]. Due to the lack of extended π-electron delocalization, purely inorganic materials usually have low NLO responses (e.g., semiconductors). They also have some drawbacks such as difficulty of synthesis, lack of optical quality, and slow electro-optic response times. Organic compounds are other potential candidates for NLO materials because of their strong NLO properties, high optical damage threshold, low cost and short response time to optical excitation. Urea, for example, is one of the most promising materials for NLO application, with SHG being 2.5 times that of ADP [7]. It has become evident that conjugated molecules having a strongly electron-donating group (D) connected to an electron-accepting unit (A) via a π-conjugated bridge often show highly nonlinear optical effects due to the D→A intramolecular charge transfer (ICT) [8]. In fact, some of them exhibit the largest known nonlinear coefficients, often considerably larger than those of their inorganic counterparts. For example, the second-order NLO coefficient of DSAT crystals is ten times as large as that of the inorganic standard $LiNbO_3$ [9]. The major advantage of using organic materials for NLO applications lies in the ease of fine-tuning of the NLO properties by rational modification of their molecular structure and functional groups. Moreover, organic compounds, in general, are easy to process and integrate into optical devices. Unfortunately, most of the organic π-conjugated molecules crystallize in centrosymmetric space groups due to favored anti-parallel stacking, therefore producing materials with no second order bulk susceptibility. They also suffer from poor physicochemical stability, low hardness, and cleavage tendency that obstruct their device applications.

Metal–organic frameworks, often called MOFs, are a new class of inorganic–organic hybrid materials that have enormous potential for many practical applications [10]. They can combine the high nonlinear optical coefficients of the organic molecules with excellent physical properties of the inorganics, and may thus be promising candidates for NLO materials. The coordination of organic ligands to metal ions can not only give rise to the metal-to-ligand charge transfer transitions, which are important for SHG, but also allow the organic chromophores to be arranged in orderly geometries, such as octahedral, tetrahedral and those that are less commonly observed in organic materials, enabling multidirectional charge transfer. Furthermore, a large number of such compounds possessing a range of network structures can be readily made by choosing appropriate molecular building blocks (typically metal ions or clusters) in combination with multitopic organic ligands. The almost unlimited combination of building blocks and organic linkers would, in principle, lead to assemble of a vast number of MOFs, allowing the large-scale screening and design of new NLO-active MOF materials. In addition, MOFs offer a large number of design possibilities and the advantage of tailorability: a fine-tuning of the NLO properties can be achieved by rational modification of the chemical composition and structure [11]. While a variety of materials including inorganic, organometallic, organic, and polymeric systems have been studied for their NLO activity, the development of NLO materials based on metal–organic frameworks still remains in its infancy [12].

2 Metal–Organic Frameworks for Second-Order Nonlinear Optics

For a material to exhibit NLO activity it should be noncentrosymmetric. Therefore, rational design and synthesis of noncentrosymmetric MOFs is the prerequisite and key issue to obtain NLO MOFs materials. A straightforward strategy for the efficient synthesis of noncentrosymmetric MOFs includes the coordination of chiral organic ligands to metal ions or the use of chiral templates to induce homochirality into the framework [13]. They can also be made from achiral components via spontaneous resolution during crystal growth without involving any enantiopure precursors [14, 15]. In some cases, unsymmetrical ligands can help remove the centrosymmetry from the frameworks, in particular those with diamondoid or grid topologies [12]. Another synthetic approach to noncentrosymmetric MOFs is based on mixed metal carboxylates. For example, it has been found that the incorporation of alkali or alkaline earth ions into the Cd^{2+}/carboxylate frameworks may greatly enhance the possibility of generating acentric MOFs, despite that the underlying causes of noncentrosymmetry have not been fully elucidated [16, 17]. By following the approaches outlined above, a large number of noncentrosymmetric MOFs have been successfully synthesized. However, in order to minimize optical losses in SHG processes, only those MOFs which are transparent in the working frequency range may have possible NLO applications. Thus, we will focus on the synthesis, structures of MOFs with d^{10} metal ions such as Zn^{2+}, Cd^{2+}, and pay more attention to those with large SHG efficiencies.

2.1 NLO MOFs Built from Chiral Ligands

The chirality associated with MOFs can ensure noncentrosymmetric solid-state organization, which suggests potential second-order nonlinear optical effects in these materials. In principle, all chiral MOFs are SHG active. However, they do not necessarily display strong SHG responses as the dipoles of chromophoric building blocks in the chiral MOFs structures do not have to orient in a noncentrosymmetric fashion. Although it has been found that the mode of metal coordination and MOF topology may influence the solid-state SHG in these chiral MOFs, the direct correlation between the structures and the observed SHG has not yet been established. Nevertheless, it has become evident that, similar to organic NLO materials, chiral MOFs with push–pull centers (with electron donor–acceptor systems) generally have large SHG responses [18].

The design and construction of chiral MOFs can be readily achieved by self-assembly of metal ions with chiral ligands, which involves the transfer of chiral information from a chiral ligand to the metal center and then to the whole framework. It should be pointed out that this is not the surest way to produce chiral MOFs, as in some cases mesomeric phenomenon may occur and the path for the transfer of chiral information may be interrupted. In this context, chiral ligands can be classified into two types: nonaromatic and aromatic chiral ligands in terms of whether they contain a conjugated system or not. Nonaromatic chiral ligands, in particular those with carboxylic groups and/or N-donor atoms have the capability to organize metal ions into chiral or acentric MOFs. For example, the reactions of *S*-carboxymethyl-L-cysteine with Zn^{2+} and Cd^{2+} ions gave two chiral MOFs $[Zn(H_2O)(SCMC)]_n$ and $\{[Cd(H_2O)(SCMC)] \cdot 2H_2O\}_n$ [19], which exhibit wave-like 2D (4,4) layers that are further assembled into a 3D framework through hydrogen bonding interactions. Both of them crystallize in the chiral space groups and show SHG intensities of 0.05 and 0.06, respectively relative to that of urea. A 3D chiral MOF $[Mn(H_2O)(SCMC)]$ constructed by Mn^{2+} and SCMC was also synthesized [20]. Structural analysis reveals that the four symmetry-related Mn^{2+} ions in this compound are bridged by four SCMC ligands through N and O atoms, generating a 3D MOF with a 6^6 topology. Upon excitation by a high-intensity Nd:YAG laser, a weak green light emission from the crystal was observed, confirming its noncentrosymmetric structure.

Some other naturally chiral carboxylate ligands were also selected to construct chiral MOFs. For example, assemblies of Zn^{2+} and Cd^{2+} with D-camphoric acid in combination with some semi-rigid N-donor auxiliary ligands, such as 1,3- and 1,4-bis(imidazol-1-ylmethyl)benzene and 4,4′-dipyridylsulfide, result in the formation of chiral MOFs $[M(1,3\text{-bimb})(D\text{-ca})]_n$ (M = Zn, Cd), $[Zn(1,4\text{-bimb})(D\text{-ca})]_n$ and a noncentrosymmetric MOF $\{[Cd_2(dpys)(D\text{-ca})_2(H_2O)_2] \cdot H_2O\}_n$, in which the 1D $Zn^{2+}(Cd^{2+})$/D-ca chains are linked by N-donor ligands to form either a 2D network or a ladder-like structure [21]. Preliminary experimental results indicate that they exhibit SHG intensities ca. 0.3, 0.1, 0.3, and 0.8 times, respectively, as large as that of urea. In fact, many MOFs with nonaromatic chiral ligands are

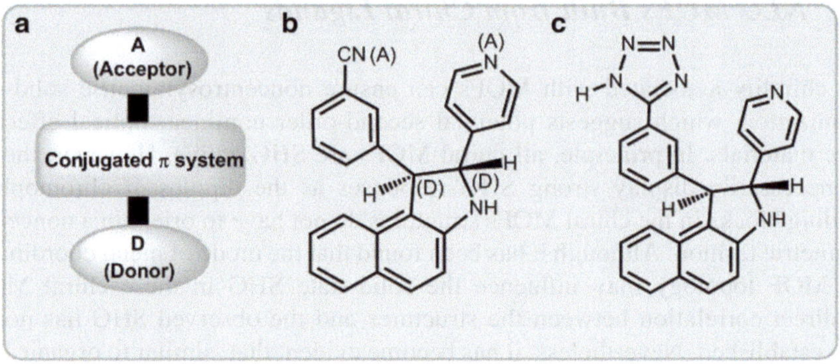

Scheme 1 (**a**) The donor-π-acceptor system in a ligand. (**b**) Schematic drawing of L showing the two-center-D–A system. (**c**) Schematic drawing of H_3tzpbin

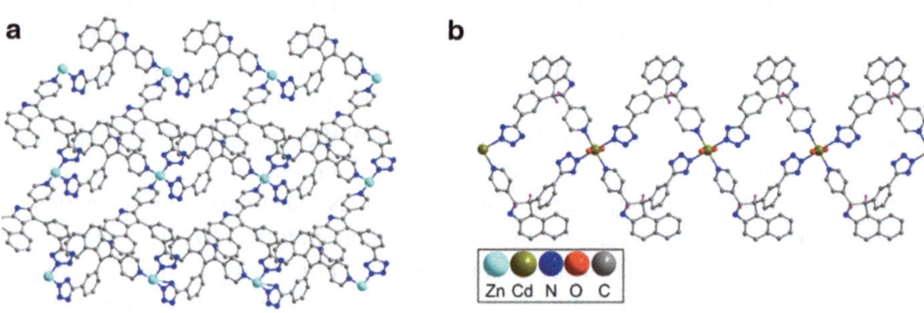

Fig. 1 (**a**) The 2D network of $[Zn(tzpbin)_2] \cdot 1.5H_2O$. After [23]. (**b**) The 1D infinite molecular box of $[Cd(H_2tzpbin)_2(H_2O)_2]_n$. After [24]

NLO active. However, they usually have weak SHG efficiencies, probably due to the lack of push–pull effect centers, an essential criterion for an SHG response.

Basically, push–pull electronic effect is more easily achieved with an aromatic chiral ligand than with a nonaromatic one as the former contains a π-conjugated system (Scheme 1a). Some chiral ligands with a push–pull character have been thus designed and employed in the reaction of metal ions, forming a range of MOFs that exhibit good SHG effect. For instance, a chiral ligand, 3-((1*R*,2*R*)-2-(pyridin-4-yl)-2,3-dihydro-1*H*-benzo[*e*]indol-1-yl)benzonitrile (L) with two chiral centers and a two-center-D–A (donor–acceptor) system has been designed (Scheme 1b) [22]. Under hydrothermal conditions, its cyano group can be readily converted into a tetrazole functionality in the presence of NaN_3, in situ generated a chiral tetrazole ligand H_3tzpbin (Scheme 1c), which subsequently reacted with Zn^{2+} and Cd^{2+} ions to afford a 2D square grid network $[Zn(tzpbin)_2] \cdot 1.5H_2O$ (Fig. 1a) and a 1D infinite molecular box $[Cd(H_2tzpbin)_2(H_2O)_2]_n$ (Fig. 1b) [23, 24]. It should be pointed out that the reaction to $[Zn(tzpbin)_2] \cdot 1.5H_2O$ was accompanied by a dehydrogenation process, making the ligand nonchiral. As the result, the product crystallized in a nonchiral space group (*F*dd2). Due to the existence of a two-center-D–A system

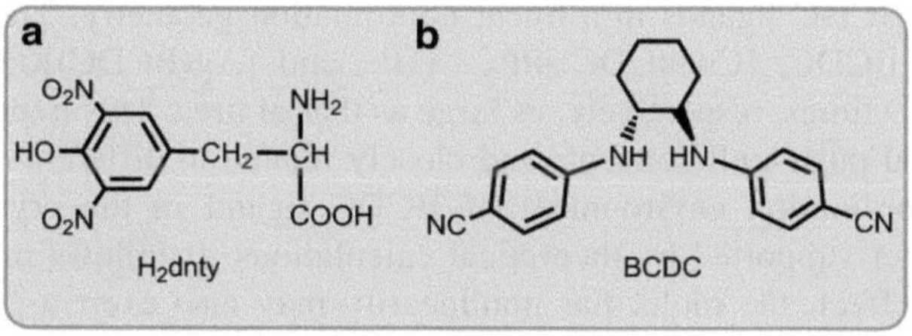

Scheme 2 Schematic drawing of H_2dnty and BCDC ligands

(multicenter push–pull electronic effect), L displays a strong SHG response, approximately 12 times larger than that of urea. In contrast, H_3tzpbin has weak SHG effect, nearly the same as that of urea. This could be due to the fact that different from 4-cyanophenyl group in L, the tetrazole ring in H_3tzpbin is not an electronic acceptor. However, when it coordinates with metal ions, donation of the lone pair of electrons from pyridyl and tetrazole nitrogen atoms onto metal center occurs, creating an excellent donor–acceptor system in [Zn(tzpbin)$_2$] · 1.5H_2O and [Cd(H_2tzpbin)$_2$(H_2O)$_2$]$_n$. As a result, they both exhibit very strong SHG responses, about 50 and 80 times, respectively, that observed for urea.

As one can see above, MOFs with a donor-π-acceptor organic chromophore ligand may have good SHG responses as the hybridization of organic chromophore and metal ions favors charge transfer in general. This strategy is not limited to closed-shell d^{10} metals. An Mn^{2+} MOF [Mn(Hdnty)$_2$] constructed by 3,5-dinitrotyrosine (Scheme 2a) has been shown to have a powder SHG efficiency of six times that of urea [25]. This compound crystallizes in a chiral space group $P2_12_12_1$ and has a 2D square-grid topological net. Although the homochiral H_2dnty chromophore itself displays a weak SHG response, once it is combined with Mn^{2+} ions, the resulting hybridized MOF can produce a large SHG effect. It appears that the coordination with metal ions may enhance the push–pull effect of the ligand, even though some adverse impact has been observed. A lanthanide compound, [Nd(Hdnty)$_2$(NO_3)(H_2O)$_5$] · 3H_2O synthesized using the same ligand has also been reported [25]. It is a 3D framework assembled through weak supramolecular interactions, such as π–π stacking and hydrogen bonds between amino and nitro groups. Similarly, this compound exhibits a strong SHG effect with an intensity of five times as great as that of urea. In this case, however, the enhancement of SHG efficiency is partially attributed to the hydrogen bonding that can also be considered as a good donor–acceptor system capable to increase charge separation.

Another interesting chiral ligand BCDC with multiple ligation sites and a push–pull center (Scheme 2b) was applied in the synthesis of NLO MOFs. Reaction of BCDC with Cu^+ and Ag^+ salts yielded chiral MOFs [Cu(BCDC)]PF_6 · THF and [Ag(BCDC)]ClO_4 [26, 27]. The copper compound is a 2D network in which the Cu^+ ions reside on a twofold axis in the crystal structure and adopt a tetrahedral environment by coordinating with two amino groups of a BCDC ligand and cyano groups of two other BCDC ligands. The silver compound has a 1D helical chain structure in which the Ag^+ ion is coordinated by two cyano nitrogen atoms from two different BCDC ligands in a linear coordination geometry. The SHG values observed for BCDC, $[Cu(BCDC)]PF_6 \cdot THF$, and $[Ag(BCDC)]ClO_4$ are about 0.3, 0.2 and 2.9 times, respectively, as large as that of urea. This trend is consistent with traditional push–pull concepts and closely related to different configurations and local coordination environment of BCDC ligand in the crystal structure, which is further supported by theoretical calculations. It follows that in addition to push–pull effect, the molecular nonlinearity may also exert a great influence on the bulk NLO property.

2.2 NLO MOFs Built from Achiral Ligands

Despite the tremendous success in the construction of NLO MOFs from chiral ligands, this synthetic strategy has suffered from the high cost and limited availability of the enantiopure chiral ligands. Thus, an alternative method by using achiral ligands is required for the development of such materials. Some chiral MOFs can be obtained by spontaneous resolution without involving any chiral components. Unfortunately, the formation of chiral products through spontaneous resolution is still unpredictable because the laws of physics determining this processes remain poorly understood up until now. Another simple but effective way to increase the chance of making noncentrosymmetric MOFs is to use unsymmetrical bridging ligands. The unsymmetrical ligands, in particular those with strong donor/acceptor substituents may, on the one hand, help remove the center of symmetry and, on the other hand, create electronic asymmetry in the MOFs network, which is the key for a high nonlinearity or substantial SHG response. Following this principle, a range of noncentrosymmetric MOFs, with 1D, 2D, and 3D network structures have been synthesized. Their SHG efficiencies vary in a wide range and appear to have no direct correlation with the types of metal ions, the ligands and the topological structures. However, it has been observed that MOFs with a 3D diamondoid structure similar to that of KDP, or with an octupolar symmetry usually exhibit large SHG activity. In addition, some 2D grids and 1D helical chains also display good SHG efficiencies.

2.2.1 NLO MOFs with 3D Diamondoid Structures

The approach to the rational design and synthesis of noncentrosymmetric MOFs by using unsymmetrical bridging ligands work best for the 3D diamondoid system. The pioneering work in this field was carried out by Lin's group. They demonstrate that a centrosymmetry breaking at a diamond net can happen when the metal tetrahedral nodes of the diamond net are bridged by unsymmetrical ligands. They first synthesized a chiral threefold interpenetrated diamondoid net, $[Zn(inic)_2]$ from $Zn(ClO_4)_2 \cdot 6H_2O$ and $inic^-$ anion (generated in situ from the hydrolysis of 4-cyanopyridine) under solvothermal conditions (Fig. 2). The tetrahedral Zn^{2+}

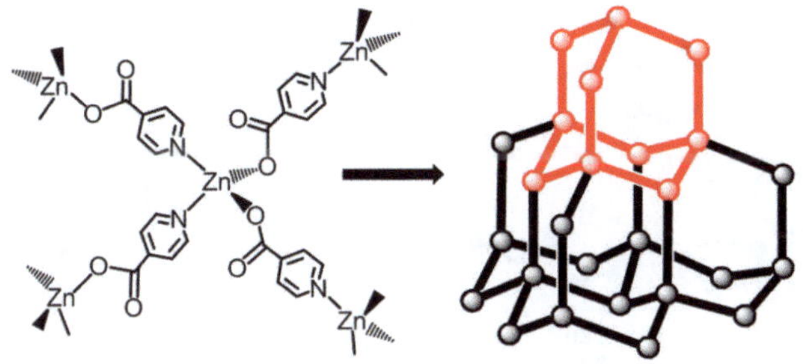

Fig. 2 A 3D diamondoid net of [Zn(inic)$_2$] built from Zn^{2+} tetrahedral connecting points and *p*-pyridinecarboxylate linkers. After [12]

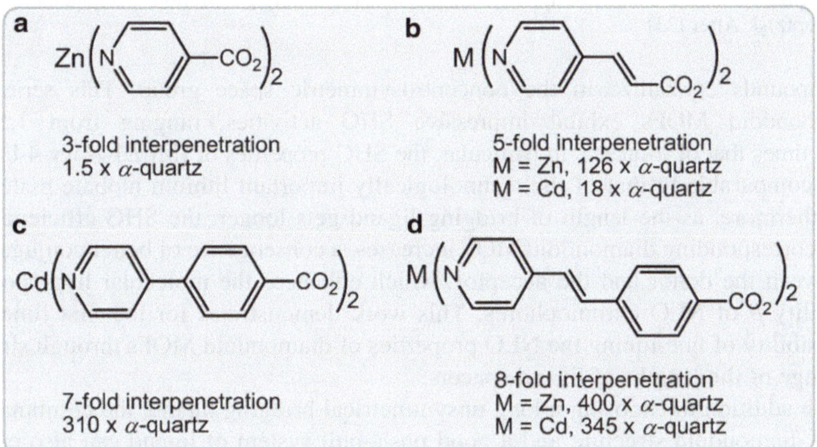

Scheme 3 3D MOFs of diamondoid topology built from Zn^{2+}/Cd^{2+} tetrahedral connecting nodes and *p*-pyridinecarboxylate type of unsymmetrical ligands

centers in this compound are connected through pyridine and carboxylate groups of unsymmetrical inic$^-$ anions to form a noncentrosymmetric threefold interpenetrated diamondoid framework [28]. This compound shows a weak SHG signal, being 1.5 times more intense than that of α-quartz (Scheme 3a).

By tuning the length of *p*-pyridinecarboxylate linkers, similar reactions between $Zn(ClO_4)_2 \cdot 6H_2O$ or $Cd(ClO_4)_2 \cdot 6H_2O$ and corresponding precursors afforded more noncentrosymmetric diamondoid MOFs with systematically elongated linear spacers. It has been found that the longer the bridging ligand, the higher the interpenetration degree. For example, the reactions with 4-pyridylacrylic acid led to diamondoid networks, [Zn(4-pya)$_2$] [28], and [Cd(4-pya)$_2$] · H_2O [29] with fivefold interpenetration (Scheme 3b), while with the longer 4-(4-pyridyl)benzoate and (*E*)-4-(2-(4-pyridyl)vinyl)benzoate afforded a sevenfold interpenetrated [Cd(pyb)$_2$] · H_2O (Scheme 3c) and eightfold interpenetrated [M((*E*)-4-pyv-4-bza)$_2$] (M = Zn, Cd) (Scheme 3d) diamondoid networks, respectively [29, 30]. As expected, all these

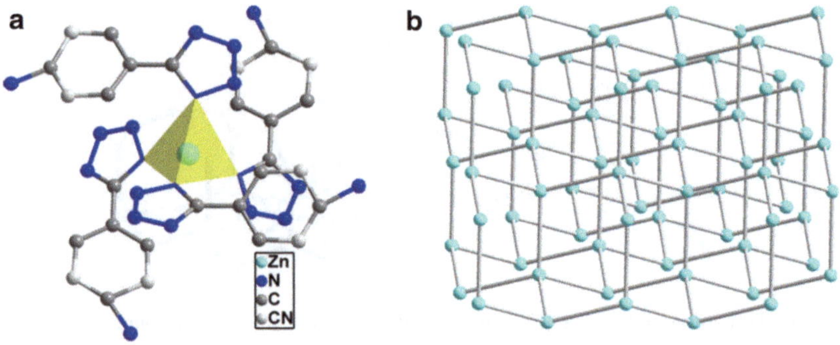

Fig. 3 (**a**) Asymmetric unit representation of [Zn(aptz)$_2$] showing that the Zn center displays a slightly distorted tetrahedron. After [23]. (**b**) A non-interpenetration diamondoid net of [Zn(aptz)$_2$]. After [23]

compounds crystallize in the noncentrosymmetric space group. This series of diamondoid MOFs exhibit impressive SHG activities, ranging from 1.5 to 400 times that of α-quartz. In particular, the SHG properties of [M((*E*)-4-pyv-4-bza)$_2$] are comparable to that of the technologically important lithium niobate material. Furthermore, as the length of bridging ligand gets longer, the SHG efficiency of the corresponding diamondoid MOF increases, a consequence of better conjugation between the donor and the acceptor, which enhances the molecular hyperpolarizability β of NLO chromophores. This work demonstrates for the first time the possibility of fine-tuning the NLO properties of diamondoid MOFs through simple change of the lengths of ligand spacers.

In addition to choosing a long unsymmetrical bridging ligand, the combination of a diamondoid structure and a good push–pull system of ligand can also effectively improve the SHG properties. For example, Xiong and co-workers [23] synthesized a non-interpenetrated diamondoid Zn-tetrazole MOF [Zn(aptz)$_2$] which showed an SHG activity ca. 5 times that of urea. In this compound, the Zn^{2+} center adopts slightly distorted tetrahedral coordination geometry, surrounded by four N atoms from four aptz ligands (Fig. 3a). The larger steric bulk of the aptz ligand together with the short separation between Zn^{2+} ions prevents the interpenetration of the individual diamondoid networks (Fig. 3b). The strong SHG efficiency of this compound is attributed to the presence of strong donor/acceptor substituents in the aptz ligand.

Besides unsymmetrical ligands, racemic chiral ligands are also useful for making noncentrosymmetric diamondoid MOFs. Racemic H-Imazethapyr ligand, for example, has been used to react with $Cd(ClO_4)_2 \cdot 6H_2O$ under hydrothermal conditions to give a diamondoid MOF, [Cd(Imazethapyr)$_2$] [31]. The Cd^{2+} ions in this MOF adopt a slightly distorted octahedral geometry, coordinated with four N atoms and two O atoms from Imazethapyr anions (Fig. 4a). Thus, each Cd^{2+} ion links four different Imazethapyr ligands in four different directions, generating a 3D framework with a diamondoid topology. No interpenetration occurs between the individual diamondoid networks due to the steric effect of the ligand and the short

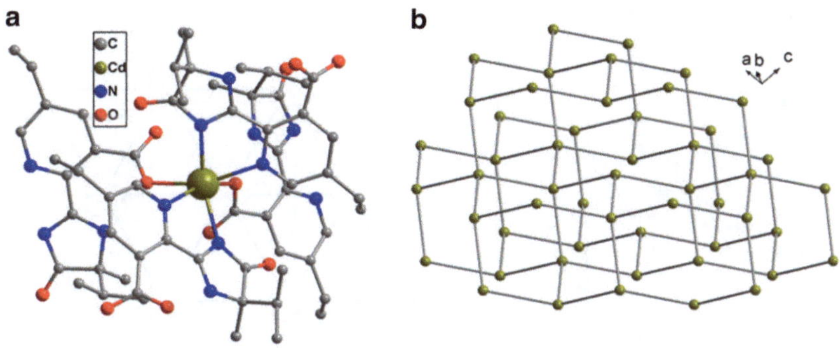

Fig. 4 (**a**) Coordination environment of Cd^{2+} in [Cd(Imazethapyr)$_2$]. After [31]. (**b**) A simplified 3D diamondoid net of [Cd(Imazethapyr)$_2$]. After [31]

distance between two adjacent Cd^{2+} ions (Fig. 4b). Preliminary studies of a powdered sample indicate that this compound is SHG active with a value ca. 20 times larger than that of KDP.

2.2.2 NLO MOFs with Octupolar Symmetries

The concept of octupolar nonlinearities was described in the early 1990 by Zyss and co-workers [32]. Due to the lack of ground state dipole moment, which is the main origin of centrosymmetric molecular packing style, octupolar chromophores show easier noncentrosymmetric crystallization in the solid state to achieve a maximum bulk effect. Besides, octupolar frameworks can be more transparent and the efficiency-transparency trade-off can be remarkably improved. Because of these advantages, increasing research attention has been focused on NLO active octupolar molecules, including organic compounds and metal coordination and organometallic complexes. Although octupolar MOFs are rarely encountered in MOFs, some of them have indeed proven to have excellent NLO properties.

A synthetic approach to noncentrosymmetric octupolar NLO materials involves using trinuclear metal carboxylates clusters as fundamental building units. In general, trinuclear metal carboxylates of group 12 metals are more difficult to obtain in comparison with those of other transition metals. Despite this, an octupolar cadmium-carboxylate MOF, formulated as [Cd_3(μ-OH)$_3$((*E*)-4-pyv-4-bza)$_6$(py)$_6$](ClO_4)$_2$, was successfully synthesized from Cd(ClO_4)$_2$ · 6H_2O and ethyl 4-[2-(4-pyridyl)ethenyl] benzoate under hydro(solvo)thermal conditions [33]. The trinuclear hydroxo-centered Cd^{2+} carboxylates, acting as a secondary building unit (SBU), possesses a threefold rotational symmetry and constitutes the octupolar NLO chromophoric unit of this compound (Fig. 5a). The SBUs are then linked via ligands to form a chiral 2D layer structure (Fig. 5b). Kurtz powder SHG measurements for this compound revealed a powder SHG intensity of 130 versus α-quartz.

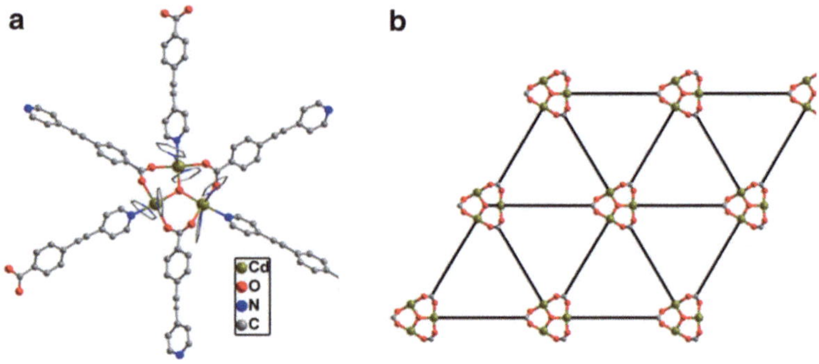

Fig. 5 (**a**) View of the cationic octupolar building block in $[Cd_3(\mu\text{-}OH)_3((E)\text{-}4\text{-pyv-}4\text{-bza})_6(py)_6](ClO_4)_2$. After [33]. (**b**) Schematic showing of the 2D sheet down from the *c* axis. After [33]

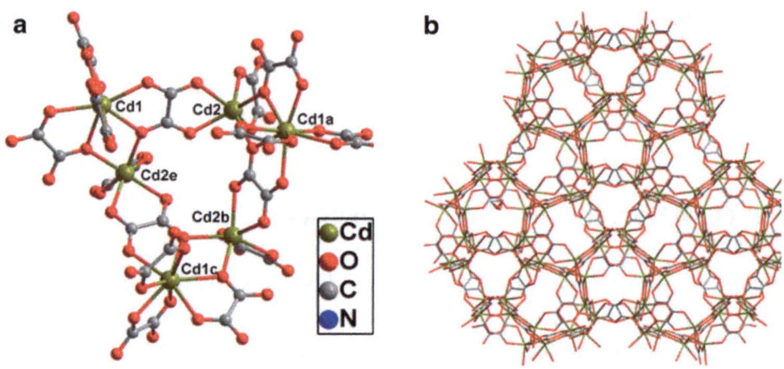

Fig. 6 (**a**) View of the hexameric octupolar building block $[Cd_6(C_2O_4)_8]^{4-}$. After [34]. (**b**) View of the octupolar 3D anionic open framework of $(H_2NMe_2)_2[Cd_3(C_2O_4)_4] \cdot MeOH \cdot 2H_2O$. After [34]

Another octupolar cadmium MOF with a large NLO efficiency ($150 \times \alpha$-quartz) has also been synthesized by a solvothermal reaction of $CdCO_3$ with symmetrical oxalic acid [34]. The product $(H_2NMe_2)_2[Cd_3(C_2O_4)_4] \cdot MeOH \cdot 2H_2O$ isolated from this reaction crystallizes in the cubic acentric space group and possesses a 3D porous coordination network. Two sets of equilateral trianglular $[Cd_3(C_2O_4)_4]^{2-}$ motifs with the center being located at a threefold axis form a bowl-shaped hexameric basic building block $[Cd_6(C_2O_4)_8]^{4-}$ (Fig. 6a). Face-to-face connection of the Cd_3 triangles forms Cd_9 cages, which undergo face-to-face intercage packing, leading to an octupolar 3D anionic supercage network (Fig. 6b) with the small and large cavities being occupied by solvent molecules and $H_2NMe_2^+$ cations, respectively. Complete replacement of $H_2NMe_2^+$ by NH_4^+, Na^+, and K^+ was accomplished and the resulting MOFs have powder SHG intensities of 155, 90, and 110, respectively, versus α-quartz, showing interesting cation-dependent SHG responses.

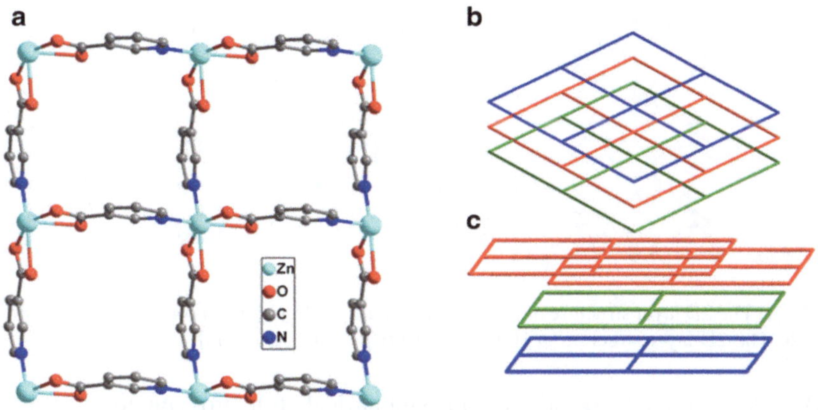

Fig. 7 (**a**) View of 2D square grid structure of [Zn(nic)$_2$]. After [35]. (**b**) A schematic showing of the interweaving of three independent rhombohedral grids in [Cd((*E*)-4-pyv-3-bza)$_2$]. After [36]. (**c**) ABC stacking of four independent 2D rhombohedral grids in [Zn$_4$((*E*)-4-pyv-3-bza)$_8$] · (H(*E*)-4-pyv-3-bza) · H_2O. After [36]

2.2.3 NLO MOFs with 2D Grid Networks

As discussed above, assemblies of Zn^{2+}/Cd^{2+} ions with unsymmetrical linear bridging ligands under hydro(solvo)thermal conditions permit the fabrication of diamondoid structures. Unfortunately, this method is not guaranteed to produce diamondoid MOFs as MOF topology is highly sensitive to small changes in the reaction conditions or the organic ligands. Sometimes 2D grid-type MOFs can be formed. In these cases, however, the use of unsymmetrical ligands can still increase the chance of obtaining noncentrosymmetric MOFs. For example, by using *m*-pyridinecarboxylate linkers instead of *p*-pyridinecarboxylate ligands, a couple of 2D grids were obtained. Similar to the diamondoid system, the elongation of ligand length results in better conjugation between the donor and the acceptor, and hence increases the NLO activity. Thus, the hydro(solvo)thermal reaction of zinc perchlorate hexahydrate with 3-cyanopyridine gave a 2D square grid [Zn(nic)$_2$], in which the Zn center has a highly distorted octahedral geometry with two carboxylate groups and two pyridyl nitrogen atoms of four different nicotinate groups in a *cis* configuration (Fig. 7a). This compound shows a low SHG effect (2 × α-quartz) [35]. Similar reactions of Cd^{2+} and Zn^{2+} salts with a longer bridging ligand 4-(3-cyanostyryl)pyridine yielded 2D rhombohedral grids [Cd((*E*)-4-pyv-3-bza)$_2$] and [Zn$_4$((*E*)-4-pyv-3-bza)$_8$] · (H(*E*)-4-pyv-3-bza) · H_2O, respectively [35, 36]. The former has a interweaving of three independent rhombohedral grids in the *ac* plane (Fig. 7b) and exhibits higher SHG intensity (1,000 × α-quartz) than the technologically important $LiNbO_3$, whereas the latter has an ABC stacking fashion of four independent 2D rhombohedral grids along the *c* axis (Fig. 7c) and also display a high SHG intensity (400 × α-quartz).

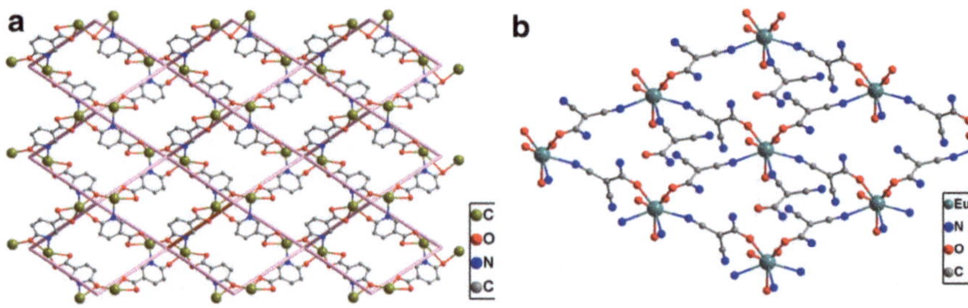

Fig. 8 (**a**) The 2D layer structure of $[Cd_8(nicO)_8(phen)_8] \cdot H_2O$ (phen is omitted for clarity). After [37]. (**b**) The 2D layer structure of $[Eu(cda)_3(H_2O)_3] \cdot H_2O$. After [38]

By using mixed ligands of unsymmetrical 6-hydroxynicotinic acid and 1,10-phenanthroline, a noncentrosymmetric grid-like MOF $[Cd_8(nicO)_8(phen)_8] \cdot H_2O$ that showed impressive SHG activity was synthesized [37]. Two distinct Cd centers, one is in a distorted mono-capped triangular prismatic coordination geometry and the other in a distorted octahedral geometry, form dinuclear building blocks which are further linked via nicO ligands to form rhombus grids extending along the *ab* plane (Fig. 8a). SHG measurement on the powder sample reveals that it exhibits an SHG efficiency of ca. 8 times that of KDP. Lanthanide metals are also capable to form 2D grids with polynitrile ligands. For example, a noncentrosymmetric 2D polynitrile Eu^{3+} MOF $[Eu(cda)_3(H_2O)_3] \cdot H_2O$ has been synthesized. The Eu^{3+} center is eight-coordinate with three nitrogen atoms from nitrile groups and five oxygen atoms: three are from water molecules and two are from amide groups (Fig. 8b). The coordination of amide-oxygen atoms to Eu^{3+} suggests a strong conjugation in cda. As a result, this compound shows strong powder SHG efficiency, being 16.8 times that of urea [38].

2.2.4 NLO MOFs with 1D Helical Chains

The systematical design of noncentrosymmetric MOFs based on 1D chains is more difficult compared to those with high-dimensional structural motifs because the 1D chain systems lack control in two other directions. Except when using enantiopure chiral ligands, most of the noncentrosymmetric linear chains (1D MOFs) are synthesized accidentally and normally have a weak SHG activity. However, some 2D and 3D MOFs constructed by helical chain motifs can display good SHG responses.

A 2D zinc MOF [Zn(spcp)(OH)] based on two types of homochiral helices was synthesized through spontaneous resolution [39]. The ligand is a sulfur-containing asymmetrical linker where sulfide moieties are well known as redox active functional groups that could enhance electronic asymmetry. The structural feature of this MOF is the alternating arrangement of the two distinct homochiral helices along the crystallographic *b* axis (Fig. 9a). One helix is formed by hydroxo bridging Zn^{2+} atoms while the other with an opposite handedness is constructed

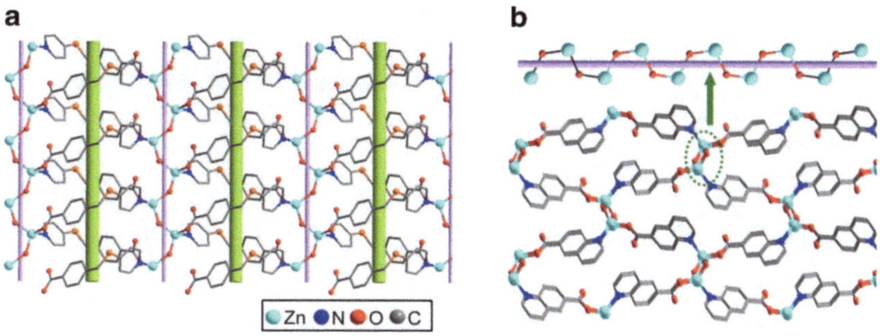

Fig. 9 (**a**) View of the two types of homochiral helices in [Zn(spcp)(OH)]. After [39]. (**b**) View of the 3D net of [Zn(OH)(quin-6-c)] along the *c* axis. After [40]

by spcp bridged between the Zn^{2+} centers. The compound crystallizes in a $P2_1$ space group and the preliminary experimental results show that it displays high powder SHG efficiency approximately five times higher than that of KDP. The large SHG response could be attributed to the big electronic asymmetry of infinite hydroxo-zinc helices. Another zinc MOF [Zn(OH)(quin-6-c)] containing similar hydroxo-zinc helices has been reported [40]. Each $[Zn(\mu\text{-}OH)_2]_n$ helix in this compound is linked to four nearest-neighbors through the bridging quin-6-c ligands along the *a* and *b* axes, generating a 3D 4-connected framework (Fig. 9b). This compound crystallizes in a noncentrosymmetric orthorhombic space group and displays a powder SHG intensity ca. 460 times in comparing with that of α-quartz.

1,2,4-Triazole and its derivatives have been widely used to design MOFs. These π-electron-rich heterocyclic ligands with versatile coordination modes are found to have great promise for constructing NLO MOFs. A typical example [41] of these compounds is $[Cd(trtr)_2]_n$ with a noncentrosymmetric polar packing arrangement. The basic structural motif of this compound is a helical chain made up on trtr ligands and Cd atoms. These right- and left-handed helices are alternatively arranged, leading to a 3D architecture that can be reduced to a binodal structure with six-connected Cd nodes and three-connected trtr nodes. It crystallizes in a noncentrosymmetric space group (*F*dd2) and displays an SHG signal six times that of KDP. The strong SHG response is mainly attributed to the good donor–acceptor system of the organic moieties which are arranged in an orderly fashion (Fig. 10a). Besides, due to the packing of the helical chains and sheets in parallel, the whole 3D crystal structure also has a dipole moment along the *c* direction which is believed to strengthen the hyperpolarizability, giving rise to a good SHG response.

Imidazole derivatives have been used with metal ions to form noncentrosymmetric MOFs as they have strong tendency to induce helical structures of axial chirality. The im-N^-–M^+ donor–acceptor systems in these MOFs are expected to enhance the hyperpolarizability β and thus the second-order NLO susceptibility. By introducing polar substituents such as – CN into the imidazole ring, the β value

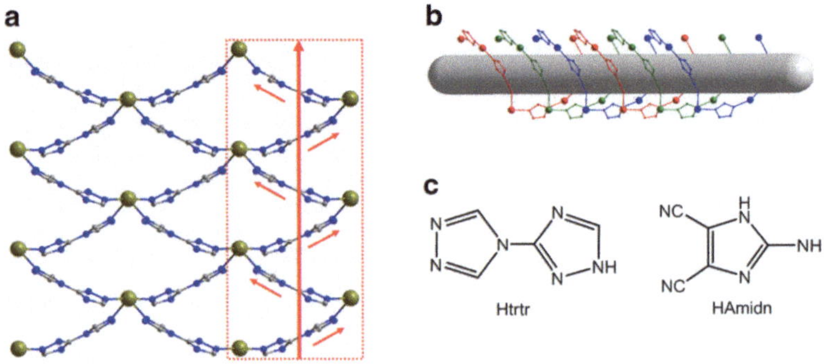

Fig. 10 (**a**) View of the 3D structure of $[Cd(trtr)_2]_n$ made up parallel right-helical chains (the *red arrows* represent dipoles of trtr ligands). After [41]. (**b**) View of the triple helices in $[Ag(Amidn)]_n$. After [42]. (**c**) Schematic drawings of Htrtr and HAmidn ligands

can be further increased. This is well demonstrated by the rigid planar imidazole ligand HAmidn which is able to combine with Ag^+ ions to form a noncentrosymmetric 3D MOF $[Ag(Amidn)]_n$ having good SHG efficiency[42]. The structure features nanotubes built of chiral Ag-Amidn triple helices which are believed to be the source of the resulting noncentrosymmetric space group (Fig. 10b). The triple right- and left-handed helices are alternately arranged to form a helix-based open framework structure. The SHG efficiency of this compound is approximately 5.4 times that of KDP. The strong SHG activity is contributed to the donor–acceptor system in which the Ag^+ ion acts as an electron donor while the imidazole ring as an electron acceptor.

2.2.5 NLO MOFs Incorporating with Alkali or Alkaline Earth Ions

In MOF synthesis, alkali or alkaline earth ions are inclined to form a cation hydration complex, acting only as a counter ion when the reaction is carried out in aqueous solution, as usually happens in the hydrothermal synthesis where water is used as a reaction medium. However, in the absence of water, they have a great opportunity to coordinate with carboxylate oxygen atoms and participate in the network construction. A number of Cd^{2+}/Li^+ carboxylate MOFs have thus been synthesized via solvothermal reactions of Cd^{2+} salt with various symmetrical aromatic polycarboxylic acids in the presence of $LiNO_3$. The main feature of these compounds is that they are all anionic frameworks in which the Cd^{2+} center is eight-coordinate, chelated by four carboxylate groups in a severely distorted mono-capped pentagonal bipyramidal geometry while the Li^+ ion acts as both a charge balancer and a connecting node to extend the network structure. The eight-coordinated Cd^{2+} centers are different from those commonly observed in the centrosymmetric Cd^{2+}/carboxylate frameworks where the octahedral Cd^{2+} centers are chelated by two carboxylate groups, leaving two mutually *cis* coordination sites to be occupied by solvent molecules or nitrogen atoms from the auxiliary ligands.

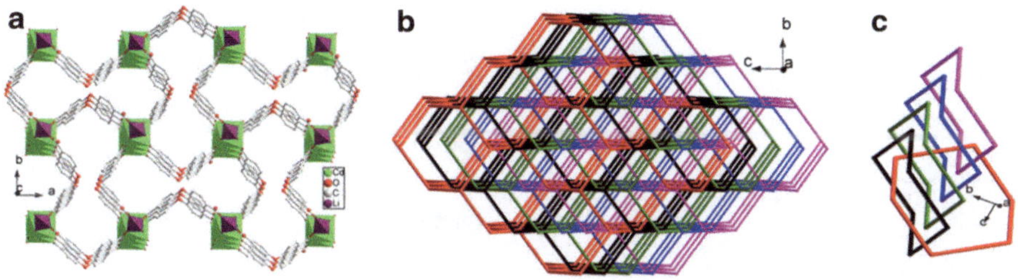

Fig. 11 (**a**) Polyhedral view of the 3D anionic framework of $[CdLi(oba)_2]^-$. (**b**) View of the Li^+-omitted fivefold interpenetrating diamondoid framework. (**c**) The catenation of the six-membered rings of five individual diamondoid nets. Reprinted with permission from [16] Copyright 2010 American Chemical Society

Despite the symmetrical ligands used, the above mixed metal system has a strong tendency to afford noncentrosymmetric structures [16], presumably due to the presence of mixed metal connecting nodes that favors noncentrosymmetric packing. For example, two noncentrosymmetric Cd^{2+}/Li^+ MOFs, $(Me_2NH_2)[CdLi(oba)_2]$ and $(Me_2NH_2)[CdLi(m\text{-bdc})_2]$ were synthesized this way. In $(Me_2NH_2)[CdLi(oba)_2]$, the Cd^{2+} center is in a distorted mono-capped pentagonal bipyramidal geometry, chelated by four carboxylate groups while the Li^+ ion is in a slightly deviated tetrahedral geometry, coordinated with four O atoms from four different carboxylate groups. The Cd^{2+} and Li^+ ions are connected through oba^{2-} ligands to form a 3D anionic framework (Fig. 11a). If the Li^+ ions are ignored, such a 3D framework can be abstracted as a typical diamondoid topological network with the Cd–Cd–Cd angles of 109.42 and 109.57°, being close to that of the perfect tetrahedral angle found in diamond (Fig. 11b). The resulting Li^+-omitted framework is a fivefold self-interpenetrated net where the six-membered rings of one diamondoid net is catenated by four six-membered rings of the other identical diamondoid nets (Fig. 11c). Thus, the overall framework of this compound can be considered as fivefold interpenetrating diamondoid nets which are linked by Li^+ connections.

The coordination environment of Cd^{2+} and Li^+ ions in $(Me_2NH_2)[CdLi(m\text{-bdc})_2]$ is similar to that in the oba^{2-} compound (Fig. 12a). The interconnections of $m\text{-bdc}^{2-}$ ligands and Cd^{2+}, together with Li^+ ions, result in a 3D anionic framework with large square channels accommodated by the $(Me_2NH_2)^+$ cations (Fig. 12b). Upon removing the Li^+ ions from the structure, the remaining anionic framework becomes a 2D square grid net with a (4,4) topology (Fig. 12c). Therefore, this compound can be viewed as being constructed by the 2D grid nets through the Cd–Li–Cd linkages. SHG measurements show that the oba^{2-} and $m\text{-bdc}^{2-}$ MOFs possess SHG intensities of five and four times that of the KDP standard, respectively. The former compound gives a slightly larger SHG intensity than the latter one, probably due to its diamondoid structure. Following this strategy, several other noncentrosymmetric Cd^{2+}/Li^+ MOFs, such as $(Me_2NH_2)[Cd(OH\text{-}m\text{-Hbdc})(OH\text{-}m\text{-bdc})] \cdot DMF \cdot CH_3OH \cdot H_2O$ and $(Me_2NH_2)[CdLi(sdba)_2] \cdot DMF \cdot CH_3OH$ have been prepared. By using alkaline earth ions instead of Li^+, a series of

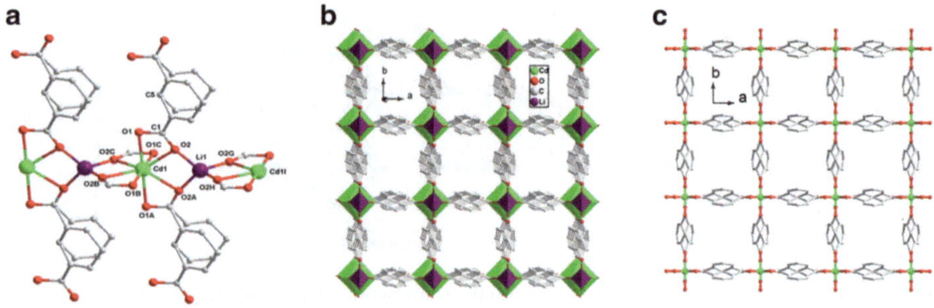

Fig. 12 (**a**) The coordination environment of Cd^{2+} and Li^{+} in $(Me_2NH_2)[CdLi(m\text{-bdc})_2]$. (**b**) Polyhedral view of the 3D anionic framework of $[CdLi(m\text{-bdc})_2]^{-}$. (**c**) View of the Li^{+}-omitted square grid in $(Me_2NH_2)[CdLi(m\text{-bdc})_2]$. Reprinted with permission from [16] Copyright 2010 American Chemical Society

Cd^{2+}/M^{2+}(M = Ca, Sr) noncentrosymmetric MOFs, i.e. $[CdCa(m\text{-bdc})_2(DMF)_2]$, $[CdCa(OH\text{-}m\text{-bdc})_2(H_2O)_2] \cdot 2Me_2NH$ and $[CdSr_2(m\text{-bdc})_2(NO_3)_2(DMF)_4]$ were obtained [17]. However, they all display weak SHG efficiencies.

3 Conclusions and Perspectives

In this chapter we briefly review the recent progresses in the design and synthesis of metal–organic frameworks as a new class of materials with potential for NLO application. The noncentrosymmetric MOFs, discussed herein are synthesized by using either chiral ligands or unsymmetrical ligands to assemble with suitable metal ions under hydro(solvo)thermal conditions. Additionally, incorporation of alkali or alkaline earth ions into Cd^{2+}/carboxylate frameworks is also an effective way to produce noncentrosymmetric MOFs.

Thanks to the strong and highly directional metal-ligand coordination bonds, as well as the virtually unlimited structural possibilities involving combinations of inorganic and organic building components, MOFs can be rationally designed and structurally tailored to fit NLO applications by judiciously choosing metal nodes and bridging organic ligands. It has been found that similar to NLO organic materials, NLO MOFs with push–pull systems usually exhibit good SHG responses. From the structural point of view, diamondoid or octupolar MOFs often show impressive SHG efficiencies. However, a more reliable method to synthesize noncentrosymmetric MOFs and a comprehensive theory for better understanding and revealing the structure-NLO property relationship in the NLO MOFs are still needed. More recently, inclusion of the ordered organic dipolar chromophore DPASD in a microporous MOF has been reported [43]. This encapsulated MOF shows a highly NLO activity, which opens up new possibilities in the area of NLO MOF materials.

Acknowledgments We acknowledge the financial supports from the National Basic Research Program of China (973 Program, 2012CB821702), the National Natural Science Foundation of China (21233009 and 21173221) and the State Key Laboratory of Structural Chemistry, Fujian Institute of Research on the Structure of Matter, Chinese Academy of Sciences.

References

1. Franken PA, Hill AE, Peters CW, Weinreich G (1961) Generation of optical harmonics. Phys Rev Lett 7:118–119
2. Saleh BEA, Teich MC (1991) Fundamentals of photonics. Wiley, New York
3. Boyd RW (1992) Nonlinear optics. Academic, San Diego
4. Zyss J (1994) Molecular nonlinear optics, materials, physics and devices. Academic, New York
5. Fan YX, Eckardt RC, Byer RL, Route RK, Feigelson RS (1984) $AgGaS_2$ infrared parametric oscillator. Appl Phys Lett 45:313–315
6. Glass AM (1988) Materials for photonic switching and Information processing. MRS Bull 13:16–20
7. Donaldson WR, Tang CL (1984) Urea optical parametric oscillator. Appl Phys Lett 44:25–27
8. Davydov BL, Derkacheva LD, Dunina VV, Zhabotinskii ME, Zolin VF, Koreneva LG, Samokhina MA (1970) Connection between charge transfer and laser second harmonic generation. JETP Lett 12:16–19
9. Marder SR, Perry JW, Schaefer WP (1989) Synthesis of organic salts with large second-order optical nonlinearities. Science 245:626–628
10. Farrusseng D (2011) Metal–organic frameworks: applications from catalysis to gas storage. Wiley-VCH Verlag & Co. KgaA, Weinheim
11. Evans OR, Lin WB (2002) Crystal engineering of NLO materials based on metal–organic coordination networks. Acc Chem Res 35:511–522
12. Wang C, Zhang T, Lin WB (2012) Rational synthesis of noncentrosymmetric metal–organic frameworks for second-order nonlinear optics. Chem Rev 112:1084–1104
13. Nickerl G, Henschel A, Grünker R, Gedrich K, Kaskel S (2011) Chiral metal–organic frameworks and their application in asymmetric catalysis and stereoselective separation. Chemie Ingenieur Technik 83:90–103
14. Lin WB (2005) Homochiral porous metal–organic framework: why and how? J Solid State Chem 178:2486–2409
15. Morris RE, Bu X (2010) Induction of chiral porous solids containing only achiral building blocks. Nat Chem 2:353–361
16. Lin JD, Long XF, Lin P, Du SW (2010) A series of cation-templated, polycarboxylate-based Cd(II) or Cd(II)/Li(I) frameworks with second-order nonlinear optical and ferroelectric properties. Cryst Growth Des 10:146–157
17. Lin JD, Wu ST, Li ZH, Du SW (2010) Syntheses, topological analyses, and NLO-active properties of new Cd(II)/M(II) ($M = Ca$, Sr) metal–organic frameworks based on R-isophthalic acids ($R = H$, OH, and *t*-Bu). Dalton Trans 39:10719–10728
18. Zhao H, Qu ZR, Ye HY, Xiong RG (2008) In situ hydrothermal synthesis of tetrazole coordination polymers with interesting physical properties. Chem Soc Rev 37:84–100
19. Wang YT, Fan HH, Wang HZ, Chen XM (2005) Homochiral helical wavelike (4,4) networks constructed by divalent metal ions and *S*-carboxymethyl-L-cysteine. J Mol Struct 740:61–67
20. Xu W, Liu W, Yao FY, Zheng YQ (2011) Synthesis, crystal structure and properties of the novel chiral 3D coordination polymer with *S*-carboxymethyl-L-cysteine. Inorg Chim Acta 365:297–301
21. Liang XQ, Li DP, Li CH, Zhou XH, Li YZ, Zuo JL, You XZ (2010) Syntheses, structures, and physical properties of camphorate coordination polymers controlled by semirigid auxiliary

ligands with variable coordination positions and conformations. Cryst Growth Des 10: 2596–2605
22. Zhao H, Li YH, Wang XS, Qu ZR, Wang LZ, Xiong RG, Abrahams BF, Xue Z (2004) Noncentrosymmetric organic solids with very strong harmonic generation response. Chem Eur J 10:2386–2390
23. Ye Q, Li YH, Song YM, Huang XF, Xiong RG, Xue Z (2005) A second-order nonlinear optical material prepared through in situ hydrothermal ligand synthesis. Inorg Chem 44:3618–3625
24. Ye Q, Tang YZ, Wang XS, Xiong RG (2005) Strong enhancement of second-harmonic generation (SHG) response through multi-chiral centers and metal-coordination. Dalton Trans 1570–1573
25. Ye Q, Li YH, Wu Q, Song YM, Wang JX, Zhao H, Xiong RG, Xue Z (2005) The first metal (Nd^{3+}, Mn^{2+}, and Pb^{2+}) coordination compounds of 3,5-dinitrotyrosine and their nonlinear optical properties. Chem Eur J 11:988–994
26. Anthony SP, Radhakrishnan TP (2004) Helical and network coordination polymers based on a novel C_2-symmetric ligand: SHG enhancement through specific metal coordination. Chem Commun 1058–1059
27. Anthony SP, Radhakrishnan TP (2004) Coordination polymers of Cu(I) with a chiral push-pull ligand: Hierarchical network structures and second harmonic generation. Cryst Growth Des 4: 1223–1227
28. Evans OR, Xiong RG, Wang Z, Wong GK, Lin W (1999) Crystal engineering of acentric diamondoid metal–organic coordination networks. Angew Chem Int Ed 38:536–538
29. Evans OR, Lin W (2001) Crystal engineering of nonlinear optical materials based on interpenetrated diamondoid coordination networks. Chem Mater 13:2705–2712
30. Lin W, Ma L, Evans OR (2000) NLO-active zinc(II) and cadmium(II) coordination networks with 8-fold diamondoid structures. Chem Commun 2263–2264
31. Fu DW, Zhang W, Xiong RG (2008) The first metal–organic framework (MOF) of imazethapyr and its SHG, piezoelectric and ferroelectric properties. Dalton Trans 3946–3948
32. Ledoux I, Zyss J, Siegel JS, Brienne J, Lehn JM (1990) Second-harmonic generation from non-dipolar non-centrosymmetric aromatic charge-transfer molecules. Chem Phys Lett 172: 440–444
33. Lin W, Wang Z, Ma L (1999) A novel octupolar metal–organic NLO mateiral based on a chiral 2D coordination network. J Am Chem Soc 121:11249–11250
34. Liu Y, Li G, Li X, Cui Y (2007) Cation-dependent nonlinear optical behavior in an octupolar 3D anionic metal–organic open framework. Angew Chem Int Ed 46:6301–6304
35. Lin W, Evans OR, Xiong RG, Wang Z (1998) Supramolecular engineering of chiral and acentric 2D networks. Synthesis, structures, and second-order nonlinear optical properties of bis(nicotinato)zinc and bis{3-[2-(4-pyridyl)ethenyl]benzoato} · cadmium. J Am Chem Soc 120:13272–13273
36. Evans OR, Lin W (2001) Rational design of nonlinear optical materials based on 2D coordination networks. Chem Mater 13:3009–3017
37. He YH, Lan YZ, Zhan CH, Feng YL, Su H (2009) A stable second-order NLO and luminescent Cd(II) complex based on 6-hydroxynicotinic acid. Inorg Chim Acta 362:1952–1956
38. Shi JM, Xu W, Liu QY, Liu FL, Huang ZL, Lei H, Yu WT, Fang Q (2002) Polynitrile-bridged two-dimensional crystal: Eu(III) complex with strong fluorescence emission and NLO property. Chem Commun 756–757
39. Han L, Hong M, Wang R, Luo J, Lin Z, Yuan D (2003) A novel nonlinear optically active tubular coordination network based on two distinct homo-chiral helices. Chem Commun 2580–2581
40. Hu S, Zou HH, Zeng MH, Wang QX, Liang H (2008) Molecular packing variation of crimpled 2D layers and 3D uncommon $6^5 \cdot 8$ topology: effect of ligand on the construction of metal-quinoline-6-carboxylate polymers. Cryst Growth Des 8:2346–2351

41. Zhou WW, Chen JT, Xu G, Wang MS, Zou JP, Long XF, Wang GJ, Guo GC, Huang JS (2008) Nonlinear optical and ferroelectric properties of a 3-D Cd(II) triazolate complex with a novel $(6^3)_2(6^{10} \cdot 8^5)$ topology. Chem Commun 2762–2764
42. Yang H, Sang RL, Xu X, Xu L (2013) An unprecedented 3-D SHG MOF material of silver(I) induced by chiral triple helices. Chem Commun 49:2909–2911
43. Yu J, Cui Y, Wu C, Yang Y, Wang Z, O'Keeffe M, Chen B, Qian G (2012) Second-order nonlinear optical activity induced by ordered dipolar chromophores confined in the pores of an anionic metal–organic framework. Angew Chem Int Ed 51:10542–10545

Struct Bond (2014) 157: 167–186
DOI: 10.1007/430_2013_106

Published online: 19 June 2013

Host–Guest Metal–Organic Frameworks for Photonics

Kenji Hirai, Paolo Falcaro, Susumu Kitagawa, and Shuhei Furukawa

Abstract The assembly of metal ions and organic linkers gives the highly regulated framework scaffolds, the so-called metal–organic framework (MOFs) or porous coordination polymers (PCPs). MOFs offer fascinating platforms in which light emitting components can be rationally incorporated. A variety of metal ions and organic linkers can be used to fabricate the MOF materials with a wide range of emissive properties. Besides their inherent luminescent properties, the permanent porosity of MOFs enables to accommodate guest species therein. The accommodation of guests in the pores results in the shift of emission wavelength, the change of emission intensity or even the generation of new emission bands. Therefore, the luminescent MOFs can be potentially exploited as a chemical sensor for small molecules or ions. In this chapter, we present a variety of luminescent properties derived from the guest accommodation in MOFs, and we discuss potential applications of luminescent MOFs as sensing materials.

K. Hirai
Department of Synthetic Chemistry and Biological Chemistry, Graduate School of Engineering, Kyoto University, Katsura, Nishikyo-ku, Kyoto 615-8510, Japan

P. Falcaro
Division of Materials Science and Engineering, CSIRO, Private Bag 33, Clayton South MDC, VIC 3169, Australia

S. Kitagawa
Department of Synthetic Chemistry and Biological Chemistry, Graduate School of Engineering, Kyoto University, Katsura, Nishikyo-ku, Kyoto 615-8510, Japan

Institute for Integrated Cell-Material Sciences, Kyoto University, Yoshida, Sakyo-ku, Kyoto 606-8501, Japan

S. Furukawa (✉)
Institute for Integrated Cell-Material Sciences, Kyoto University, Yoshida, Sakyo-ku, Kyoto 606-8501, Japan
e-mail: shuhei.furukawa@icems.kyoto-u.ac.jp

Keywords Charge transfer · Chemical sensor · Energy transfer · Metal–organic frameworks · Porous coordination polymers · Structural transformation

Contents

Abbreviations

1,2,4-BTC	Benzene-1,2,4-tricarboxylate
2,4-DNP	2,4-Dinitrophenol
ad	Adeninate
adc	9,10-Anthracenedicarboxylate
bdc	1,4-Benzendicarboxylate
bpdc	Biphenyldicarboxylate
bpdc	4,4′-Biphenyldicarboxylate
BPT	Biphenyl-3,4′,5-tricarboxylate
btc	1,2,4,5-Benzenetetracarboxylate
CT	Charge transfer
dabco	1,4-Diazabicyclo[2.2.2]octane
DMA	*N*,*N*-dimethylaniline
DMF	*N*,*N*-dimethylformamide
DMPT	*N*,*N*-dimethyl-p-toluidine
dpndi	*N*,*N*′-di(4-pyridyl)-1,4,5,8-naphthalenetetracarboxydiimide
DSB	Distrylbenzene
DXP	*N*,*N*-bis(2,6-dimethylphenyl)-3,4:9,10-perylene tetracarboxylic diimide
FIrpic	(2-Carboxypyridyl)bis(3,5-difluoro-2-(2-pyridyl)phenyl)iridium(III)
H_2dtoa	Dithiooxamide
IDC	Imidazole-4,5-dicarboxylate
JAST	Jungle-gym analogue structure
MA	*N*-methylaniline
MIL	Matériaux de l’Institut Lavoisier
MOF	Metal–organic framework

MT	5-Methyl-1*H*-tetrazole
NDC	2,6-Napthalenedicarboxylate
NP	4-Nitrophenolbdc
OBA	4,4′-Oxybis(benzoate)
ox	Oxalate
PCA	4-Pyridinecaboxylate
PCP	Porous coordination polymer
PDA	Pyridine-2,6-dicarboxylic acid
pdc	Pyridine-3,5-dicarboxylate
ppy	2-Phenylpyridine
QD	Quantum dot
Rh6G	Rhodamine 6G
TBAPy	1,3,6,8-Tetrakis(benzoic acid)pyrene
TNP	2,4,6-Trinitrophenol
UMCM	University of Michigan Crystalline Material

1 Introduction

Metal–organic frameworks (MOFs) or porous coordination polymers (PCPs) are an intriguing class of crystalline materials owing to a wide variety of potential applications such as adsorption, separation, and catalysis. To date, their excellent properties related to gas separation and storage have been intensively studied; however, the research on MOFs is currently expanding towards a new application wherein their porous and physical properties are synergistically linked. In particular, MOFs offer fascinating platforms for the development of solid-state luminescent materials, because the light-emissive building blocks can be rationally incorporated into the framework scaffolds. The introduction of a variety of π-conjugated ligands and luminescent metal ions allows for the design of a wide range of emissive phenomena. Besides the inherent luminescent properties of MOFs, the capability to accommodate guest species in the pores leads to luminescent features that have not yet been observed in other solid-state materials [1, 2].

The luminescent properties of MOFs would take an advantage in a number of different ways of the ability to encapsulate guest species within the framework. For instance, even non-luminescent MOFs can be additionally endowed with the luminescent properties by encapsulating luminophores [3]. By designing the size, shape, and chemical functionality of pores, luminophores such as dye molecules, lanthanide ions, and quantum dots can be precisely encapsulated within the porous framework; this fabrication method provides an efficient protocol for the introduction of desired luminescent properties into the MOFs. In addition, the luminescent properties of the photoactive sites of MOFs are strongly affected by surrounding guest species. The accommodation of guest species usually results in the shift of emission wavelength or the change of emission intensity. In other case, a new emission band can be observed as a consequence of excimer or exciplex formation.

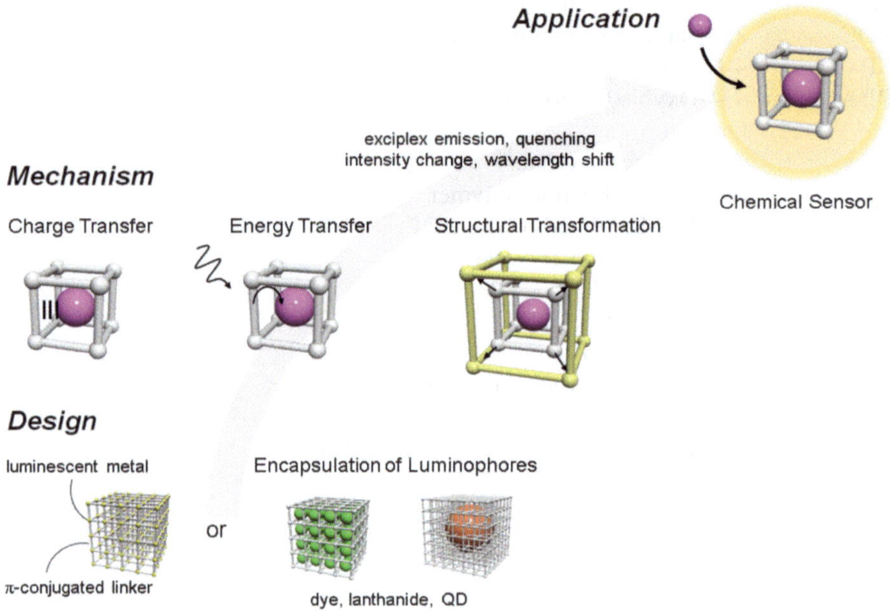

Fig. 1 Host–guest chemistry in luminescent MOFs

The mechanisms imparting luminescent functionalities are broadly classified into three categories: charge transfer (CT), energy transfer, and structural change.

1. The charge transfer between the organic linkers and guest species leads to the formation of exciplex state or results in luminescence quenching. In particular, guest species with nitro or thiol groups deprive the luminescent properties of MOFs and quench their luminescence.
2. The energy transfer between host frameworks and guest species provides a new pathway across the excitation and emission processes, which results in the shift of emission wavelength, luminescence quenching or antenna effect.
3. The guest accommodation causes the structural transformation around the photoactive centers of MOFs, which induces a difference in the luminescent intensity. Furthermore, the guest species included in the pores often restrict the structural vibration of the host frameworks and reduce the quenching effect on the luminescence.

These characteristic features allow the exploitation of luminescent MOFs as chemical sensors. The rational design of MOFs for specific sensing applications and the investigation of their luminescent response is an emerging research field. The purpose of this chapter is to present an overview of the recent development on host–guest chemistry for the preparation of luminescent MOFs (Fig. 1).

2 Guest-Directed Design of Luminescent MOFs

The unique host–guest chemistry of MOFs can be utilized to implement additional properties by loading the cavities with functional molecules or metal nanoparticles. The incorporation of luminescent guest species such as lanthanide ions and fluorescent dyes into the pores of MOFs is one strategy to fabricate the hybrid luminescent materials. The large cavities that accommodate dye molecules can be synthesized by a judicious choice of the metal ions and organic linkers. Qiu et al. have successfully synthesized a new MOF with the large one-dimensional channels of 24.5 × 27.9 Å ($[Cd_3(bpdc)_3(DMF)] \cdot 5DMF \cdot 18H_2O$; bpdc = 4,4′-biphenyldicarboxylate, DMF = *N,N*-dimethylformamide). The large pores of MOF accommodated rhodamine 6G (Rh6G) dye molecules and formed a $[Cd_3(bpdc)_3(DMF)] \cdot$ Rh6G complex. The emission spectra of the $[Cd_3(bpdc)_3(DMF)] \cdot$ Rh6G complex were recorded at different temperatures (298–77 K). The peak position of $[Cd_3(bpdc)_3(DMF)] \cdot$ Rh6G complex at 563 nm remained unchanged with decreasing the temperature, while the intensity of the signal was enhanced linearly; these features reveal that $[Cd_3(bpdc)_3(DMF)] \cdot$ Rh6G is a good candidate for applications in temperature-sensing devices [4]. Not only the dye molecules but also the coordination complex can be incorporated in the large pores of MOFs. Fischer et al. reported the gas-phase loading highly emissive perylene derivative (DXP) or an iridium complex (FIrpic), into various types of MOFs (MOF-5, MOF-177, UMCM-1, and MIL-53(Al)) (DXP = *N,N*-bis(2,6-dimethylphenyl)-3,4:9,10-perylene tetracarboxylic diimide, FIrpic = (2-carboxypyridyl)bis(3,5-difluoro-2-(2-pyridyl)phenyl)iridium(III)) [5]. The resulting host–guest hybrid materials showed strong emission endowed by the guest species.

The approach based on lanthanide(III)-dopant is a successful strategy to impart luminescent properties into MOFs. Luo et al. have synthesized an MOF of $\{[NH_4]_2[Zn(btc)] \cdot 6H_2O\}_n$ (btc = 1,2,4,5-benzenetetracarboxylate) [6]. The anionic microporous framework accommodates $[NH_4]^+$ as a counter cations. The lanthanide ions can be introduced in the micropores by the replacement with the counter cation of $[NH_4]^+$. The MOF was immersed in the solution of $EuCl_3$ or $Tb(ClO_4)_3$ and successfully accommodated Eu^{3+} or Tb^{3+} in the pores (denoted as Eu@MOF and Tb@MOF). The lanthanide-doped MOFs were used as cation sensors. Such lanthanide-doped MOFs were subsequently soaked in the solution of MCl_x and some of the lanthanide ions were exchanged with the second metal ions (M = Na^+, K^+, Zn^{2+}, Ni^{2+}, Mn^{2+}, Co^{2+}, Cu^{2+}). The luminescent intensity of the metal-ion-infused M-Eu@MOF or M-Tb@MOF was highly dependent on the nature of metal ion: Na^+, K^+, Zn^{2+} ions showed a negligible effect on the luminescence intensity, whereas others showed a range of quenching effects on the luminescence intensity. In particular, Cu^{2+} and Co^{2+} ions afforded a significant quenching effect on the lanthanide-doped MOF. This study indicates that the lanthanide-doped MOFs such as Eu@MOF or Tb@MOF are suitable for highly selective, sensitive, and low-detection-limit

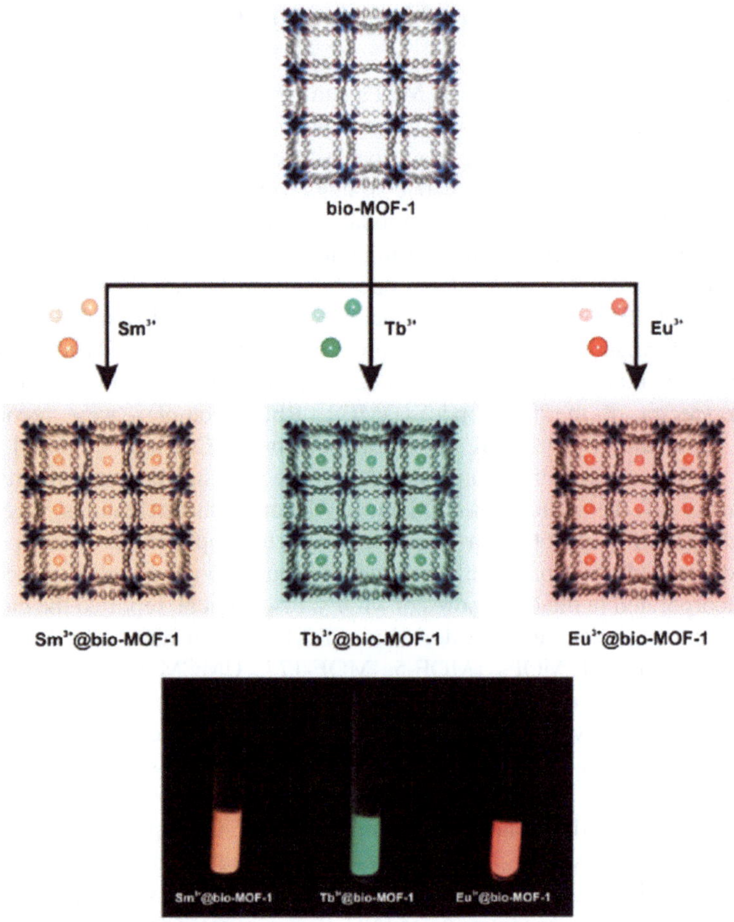

Fig. 2 Bio-MOF encapsulation and sensitization of lanthanide cations. (Reprinted from [7]. Copyright 2013 American Chemical Society)

luminescent sensors of aqueous Cu^{2+} or Co^{2+} ions. Petoud and Rosi et al. also reported a series of lanthanide-doped MOFs via cation exchange process [7]. The as-synthesized $\{[Zn_8(ad)_4(bpdc)_6O] \cdot 2Me_2NH_2 \cdot 8DMF \cdot 11H_2O\}_n$ (denoted as bio-MOF) accommodated $[Me_2NH_2]^+$ as a counter cations (ad = adeninate, bpdc = biphenyldicarboxylate). The lanthanide ions can be loaded into the pores of bio-MOF by the replacement of $[Me_2NH_2]^+$ with specific cations. The bio-MOF was soaked in the solution of $Tb(NO_3)_3$, $Sm(NO_3)_3$, $Eu(NO_3)_3$, and the lanthanide ions were successfully encapsulated. When excited at 365 nm, the doped MOFs emitted their distinctive colors (Eu^{3+}, red; Tb^{3+}, green; Sm^{3+}, orange-pink), which were detectable with the naked eye (Fig. 2). Despite the strong quenching effect of water molecules,

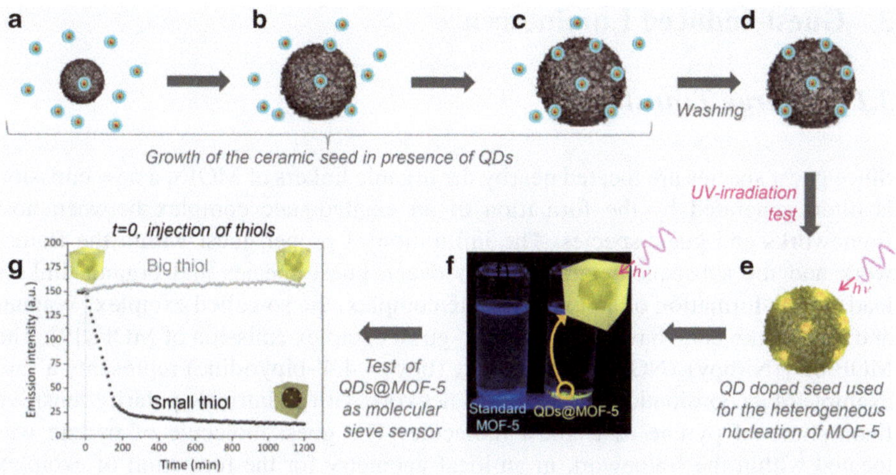

Fig. 3 Schematic of the preparation of the QDs@MOF-5 using the ceramic seeds. The seeds (α-hopeite microparticles) are grown within a colloidal solution of QDs. (**a–c**) During the QDs are embedded during the seed growth. (**d**) Subsequently the seeds are washed to remove the excess of QDs and tested with a UV lamp (365 nm) to check the presence of the luminescent particles (**e**). (**f**) The QD doped seeds are used for the heterogeneous nucleation of MOF-5; the luminescent QDs@MOFs are then washed and used as molecular sieve sensor for small thiol molecules. (**g**) The different effect on the luminescent properties using two different thiols (large thiol = *n*-isopropyl acrylamide/acrylic acid/*t*-butyl acrylamide mercaptane; small thiol = ethanethiol) along the time is presented

the characteristic luminescence of lanthanide ions was able to be observed even in aqueous environments; this experimental evidence indicates that the bio-MOF framework sufficiently protects the lanthanide ions.

A different method was proposed by Falcaro et al. [8]; the proposed protocol take advantage of ceramic micro-particle (α-hopetite) used as seeds for the nucleation of MOFs. The α-hopeite microparticles were grown in a colloidal solution of functional nanoparticles originating doped seeds. In particular a class of efficient emitting nanoparticles (quantum dots, QDs) have been embedded within the α-hopeite microparticles; subsequently, the doped-ceramic particles were used for the nucleation of MOF-5 crystals. With this method, the fabrication of hybrid materials of QD@MOF-5 with a core-shell structure was obtained (Fig. 3). The combination of functional materials was used as a molecular sieve sensor. In fact, while in a pure QDs colloidal solution the presence of any thiols affects the QDs' luminescent properties (strong luminescent emission decay towards a quenching mechanism). With the proposed core-shell system, only small thiols were able to diffuse within the MOF-5 frameworks quenching the QDs luminescence.

3 Guest-Induced Luminescence

3.1 Charge Transfer

Since guest species are located nearby the organic linkers of MOFs, a new emission is often generated by the formation of an excited-state complex between host frameworks and guest species. The infiltration of proper guest within the framework and the subsequent interaction between guest species and organic linkers leads to the formation of an excited-state complex, the so-called exciplex. Wagner and Zaworotko et al. have reported host–guest exciplex emission of MOFs [9]. The MOF of $\{[Ni(bpy)_2(NO_3)_2] \cdot 2pyrene\}_n$ (bpy = 4,4′-bipyridine) represents a rare example of a coordination polymer host network with an intrinsic polarity sensitive fluorophore of pyrene as a guest molecule. The guest molecule of pyrene was located within the framework in an ideal geometry for the formation of exciplex between bpy and pyrene. A strong fluorescence band was observed in the pyrene excimer region, suggesting that the pyrene molecules interact with bpy molecules.

MOFs are superior platforms to host π-conjugated organic species; taking advantage of this property, a photo-excited complex with guest species can be easily prepared. The π-conjugated molecules play an important role in facilitating charge or electron transfer processes. Kitagawa et al. synthesized MOFs with a π-conjugated organic linkers, ($[Zn_2(adc)_2(dabco)]_n$, denoted as JAST-6), to achieve a charge transfer (CT) interaction in the excited state between a host framework and guest molecules (adc = 9,10-anthracenedicarboxylate, dabco = 1,4-diazabicyclo[2.2.2] octane) [10]. The anthracene derivatives are suitable molecular units for MOFs because of the unique π–π* electron transitions that provide photoluminescent properties. The CT interactions between anthracene and *N*,*N*-dimethylaniline (DMA) at the excited state are well known and the case was applied to the MOF of $[Zn_2(adc)_2(dabco)]_n$. The solid-state emission spectrum of the empty state of $[Zn_2(adc)_2(dabco)]_n$ displayed an emission maximum at 415 nm, with a vibrational band at 436 nm, which can be assigned to the emission from the monomeric anthracene. After accommodating DMA in the pores, efficient quenching occurred in the excited-state fluorescence of monomeric anthracene and a new broad emission band in the 400–700 nm range was detected. The emission with large Stokes shift represents a photo-induced CT complex, an exciplex between the host anthracene unit (electron accepter) and the guest DMA molecules (electron donor). The wavelength of exciplex emission strongly depends on the ionization potential and the electron affinity of the donor and acceptor molecules. In fact, $[Zn_2(adc)_2(dabco)]_n$ adsorbed *N*-methylaniline (MA) or *N*,*N*-dimethyl-*p*-toluidine (DMPT) providing blue-shifted or red-shifted exciplex emissions (Fig. 4).

In general, strong electron withdrawing groups of the guest species often result in luminescence quenching. Chen et al. reported quenching effect of nitrobenzene on the MOF of $[Zn_4(OH)_2(1,2,4\text{-}BTC)_2]_n$ (1,2,4-BTC = benzene-1,2,4-tricarboxylate). The luminescence of the MOF was quenched by nitrobenzene molecules inside the pores. The quenching effect is attributed to the charge transfer electron transitions

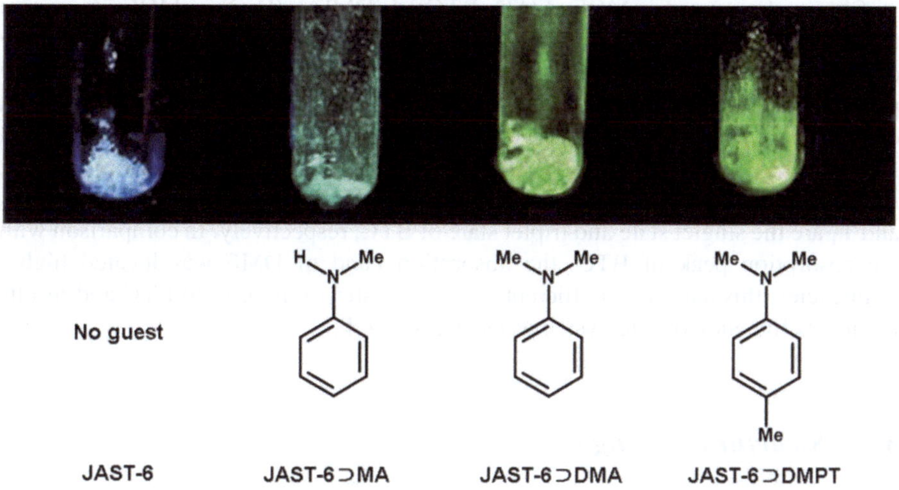

Fig. 4 A snapshot of JAST-6, JAST-6⊃MA, JAST-6⊃DMA, and JAST-6⊃DMPT. The samples were subjected to UV irradiation. (Reprinted from [10]. Copyright 2013 Royal Society of Chemistry)

from the benzene ring of BTC ligands to nitrobenzene due to the electron-deficient property of nitrobenzene and the π–π interactions between the framework and nitrobenzene molecules [11].

3.2 Energy Transfer

The energy transfer between the host frameworks and guest species often results in the shift of emission wavelength, luminescence quenching, or antenna effect. Uvdal et al. have synthesized an MOF comprising of π-conjugated dicarboxylate linkers (**L**) and a series of lanthanide metal ions (Gd(III), Eu(III) and Yb(III)) [12, 13]. Because the local environment around Ln(III) will be negatively charged due to the binding of the acetate anion to the lanthanide ion, the positively charged dye molecule of trans-4-styryl-1-methylpyridiniumiodide (**D**) was encapsulated in the MOF. The **D**-loaded MOF exhibited an efficient light-harvesting effect. The overlap between the emission spectrum of **L** itself (donor) and the absorption of **D** (acceptor) favors the Förster-type energy transfer between them [14]. The emission from the **D** in the MOF was strongly enhanced compared with direct excitation of the **D** dye. Furthermore, the delay in fluorescent decay curves was observed in time-resolved fluorescence spectroscopy. Therefore, **D** was not directly excited by the laser pulse, suggesting the energy transfer from **L** to **D**. The accommodation of dye molecules in the frameworks based on such π-conjugated linkers allows for a charge transfer couple between the organic linkers and guest species, thus leading to the highly efficient light harvesting system.

Qian et al. reported an MOF of $\{[Tb(BTC)] \cdot (DMF) \cdot (H_2O)\}_n$ (MOF-76 · guest) for the sensing of DMF and acetone [15]. The luminescent intensity of partially desolvated MOF in *n*-propanol increased and decreased with the additions of DMF and acetone solvents, respectively; remarkably, the luminescent intensity did not change in the cases of other solvents. Both absorption and luminescent spectra showed that the energy absorbed by BTC is transferred to acetone molecules, resulting in decreasing the efficiency of intersystem crossing $S_1 \rightarrow T_1$, where S_1 and T_1 are the singlet state and triplet state of BTC, respectively. In comparison with the absorption peak of BTC, the absorption band of DMF was located higher frequencies; this leads to an efficient energy transfer from DMF to BTC and results in larger efficiency of intersystem crossing $S_1 \rightarrow T_1$.

3.3 Structural Change

In general, small structural differences around the luminescent centers give significant change in the emission properties. The guest accommodation often results in the structural transformation of MOFs, thereby leading to the shift of emission wavelength and/or intensity changes. Duan et al. reported an MOF of $\{[Cu_6L_6] \cdot (H_2O)(DMSO)\}_n$, which exhibited highly selective absorption for aromatic molecules in water and different fluorescent quenching effect for toluene, nitrobenzene, aniline, and *o*-, *p*-, and *m*-dimethylbenzene (HL = 5,6-diphenyl-1,2,4-triazine-3-thiol) [16]. The luminescent spectrum of the desolvated MOF exhibited a cluster-centered band at 660 nm because of the luminescent properties of Cu_6S_6 clusters. Such critical photophysical properties were highly dependent on the Cu–Cu distance within the cluster; therefore, the emission intensities could be modified by changing the intracluster Cu–Cu distance by the inclusion of interacting guests. The accommodation of aromatic molecules within the MOF led to the system exhibiting no obvious luminescence, indicating that aromatic molecules quenched the luminescence of the cluster. The monosubstituted benzene derivatives, toluene, nitrobenzene, aniline in the pores of MOF reduced the luminescent intensity to a 25% of its original amplitude. Thus, the characteristics of substituent groups are not main factors for quenching the luminescence. Furthermore, *o*-, *p*-, and *m*-dimethylbenzene gave different luminescent intensities. The molecular structures of monosubstituted benzene and *o*-dimethylbenzene are suitable for the formation of π–π stacking between the host frameworks and guest species. As a result, the Cu–Cu distance was enlarged and the cluster-centered luminescence of Cu_6L_6 was quenched effectively.

The luminescence of lanthanide coordination compounds is generated through a three steps antenna effect. Firstly, light is absorbed by the organic ligands; secondly, the energy is transferred to Ln^{III}; finally, the Ln^{III} emits the luminescence. The vibration of organic linkers reduces the efficiency of the energy transfer from the ligand to Ln^{III}. Lu et al. reported the guest-induced intensity enhancement of lanthanide-based MOF. The lanthanide-based MOF of $K_5[Tb_5(IDC)_4(ox)_4]$

accommodated metal cations by the exchange of K^+ ion (IDC = imidazole-4,5-dicarboxylate, ox = oxalate) [17]. The intensity of the $^5D_4 \rightarrow {}^7F_5$ emission of Tb^{3+} is significantly enhanced by the addition of Ca^{2+} ions. Furthermore, the fluorescence lifetime increased from 158.90 to 287.48 μs as a consequence of the addition of Ca^{2+}. The proper amount of Ca^{2+} bonded to ox^{2-} and the interactions between Ca^{2+} and ox^{2-} suppresses the vibration of ox^{2-}, thus increasing the efficiency of the energy transfer from the ligand to Tb^{III}. Chen et al. recently demonstrated the sensing property of MOF for Cu^{2+} ion in the simulated physiological aqueous solution [18].

4 Applications

4.1 *Chemical Sensors*

Molecular recognition with the ability to distinguish among small structural differences of molecules is a critical and challenging task for a variety of applicative fields such as biotechnology, medicine, industrial processing, and environmental monitoring. In particular, the sequential system of recognition-transduction to detect these differences is a key to fabricate sensor materials for explosive, harmful, or pollutant small molecules. Most chemical sensors have been developed by following this key principle: recognition of specific molecules and transduction to convert the recognition event into a detectable signal. Luminescent MOFs are one of the promising materials to meet the requirement of this key principle. This is because the event of guest accommodation gives the differences in the luminescent emission: the shift of emission wavelength, the change of emission intensity, or the emergence of new emission as a result of exciplex formation. Therefore, MOFs can be used as chemical sensing platform to identify the specific guest molecules. A remarkable advantage of MOFs as chemical sensors relies on the designability of luminescent property by a proper choice of components. The combination of the inherent porosity and tunable luminescent properties of MOFs provides a fascinating opportunity for the fabrication of an efficient class of sensing materials. Herein, we present the sensing ability of MOFs for metal ions, organic molecules, and gas molecules.

4.2 *Sensing of Ions*

Distinguishing and detecting metal ions is the key challenges in particular for environmental monitoring and biotechnology. One way to detect specific metal ions can be implemented by the introduction of interactive sites for the target ions in MOFs.

Liu et al. first reported the cation sensing by the lanthanide-based MOF of $\{Na[EuL(H_2O)_4] \cdot 2H_2O\}_n$ (L = 1,4,8,11-tetraazacyclotetradecane-1,4,8,11-tetrapropionate) [19]. The effects of a variety of cations (Cu^{2+}, Ag^+, Zn^{2+}, Cd^{2+}, and Hg^{2+}) on the fluorescence spectrum of the MOF were investigated. The cations can diffuse within the framework and occupy the empty coordination site in the azacycle of the ligand (L). The cations of Cu^{2+}, Zn^{2+}, Cd^{2+}, and Hg^{2+} quenched the fluorescence of the Eu-based MOF. In contrast, Ag^+ gave two effects: the enhancement of the emission and the change of the emission spectrum of Eu^{3+} from multiple peaks to a single peak. The emission of $^5D_0 \rightarrow {}^7F_2$ transition was enhanced, but other emissions of $^5D_0 \rightarrow {}^7F_J$ transition ($J = 1, 4$) was diminished. This is probably because the accommodation of Ag^+ in azacycle moiety of the ligand enhanced the rigidity and modified the paramagnetic spin state of the MOF. Chen et al. also synthesized the MOF of $\{[Eu(pdc)_{1.5}] \cdot (DMF)\}_n$, comprising of the interaction site for cations (pdc = pyridine-3,5-dicarboxylate) [20]. The metal ions were captured by the pyridine moiety of pdc. While alkali and alkaline earth metal ions have a negligible effect on the luminescence intensity, Co^{2+} and Cu^{2+} significantly reduced the luminescence intensity. As an explanation for the quenching of the luminescence, Chen et al. suggest that the binding of the pyridyl nitrogen atom to Cu^{2+} or Co^{2+} reduces the efficiency of antenna effect of pdc to magnify the f–f transitions of Eu^{3+}.

It is known that the luminescent intensity of lanthanide ions strongly depends on the efficiency of energy transfer from the ligand to lanthanide ions. Cheng et al. reported the ion sensing by the lanthanide-based MOF of $[Eu(PDA)_3Mn_{1.5}(H_2O)_3]_n$ and $[Tb(PDA)_3Mn_{1.5}(H_2O)_3]_n$ (PDA = pyridine-2,6-dicarboxylic acid) [21]. The luminescent intensity of the MOFs was strongly enhanced by the addition of Zn^{2+} ions; however, no change was detected by the addition of Ca^{2+} and Mg^{2+} ions. While the addition of Mn^{2+} ions decreased the luminescent intensity, Fe^{2+}, Co^{2+}, and Ni^{2+} quenched the luminescence. The authors speculated that the addition of metal ions changed the photo-induced electron transfer process from the ligand to the Eu^{3+} or Tb^{3+} ions, thus resulting in the variety of effects of ions on the luminescent intensity. Cheng et al. also reported an Ln-based MOF of $[Eu(PDA)_3Fe_{1.5}(H_2O)_3]_n$ for metal ions sensing. Unlike the previous report, the presence of Zn^{2+} decreased the luminescent intensity of the $^5D_0 \rightarrow {}^7F_2$ transition while Mg^{2+} increased the emission intensity [22].

A number of studies on the anion sensing by luminescent MOFs have been also reported. Wong et al. synthesized an MOF of $[Tb(Mucicate)_{1.5} \cdot 3(H_2O)_2]_n$, which serves as a luminescent receptor for selective anion monitoring [23]. The luminescent intensity of the Tb^{3+} was changed by immersing MOFs in an aqueous solution of sodium salts. Anions such as I^-, CO_3^{2-}, CN^- increased the luminescent intensity, while SO_4^{2-} and PO_4^{3-} did not show any effect. The response of the channels to anions is probably related to the numerous OH groups in the mucic acid; the oxydril groups would possibly strongly interact with the anions through hydrogen bonding. The strong interaction between OH group and anions suppresses the quenching effect of OH group on the luminescence of lanthanide. Chen et al. demonstrated another method to detect anions by the luminescent change of lanthanide-based MOF [24]. MOF-76b of $\{[Tb(BTC)_3] \cdot (methanol)\}_n$ showed

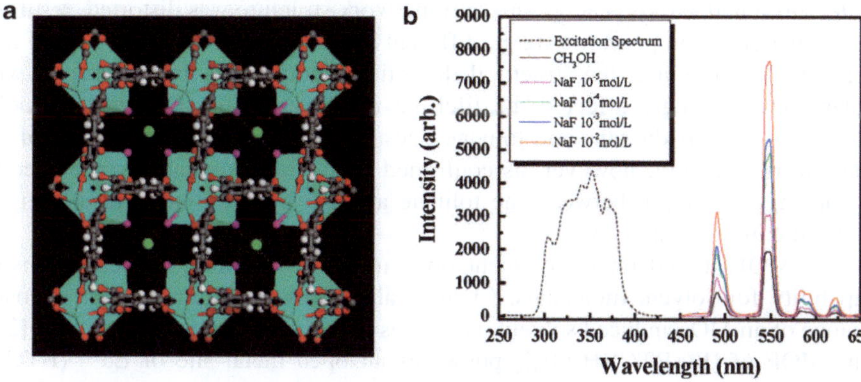

Fig. 5 (**a**) Single crystal structure of MOF-76b activated in methanol containing NaF with the model of fluoride at the center of the channel involving its hydrogen-bonding interaction with terminal methanol molecules (methanol oxygen, purple; the methyl group from methanol is omitted for clarity). (**b**) Excitation (*dotted*) and PL spectra (*solid*) of MOF-76b solid activated in different concentrations of NaF methanol solution (excited and monitored at 353 and 548 nm, respectively). (Reprinted from [24]. Copyright 2013 American Chemical Society)

the luminescent enhancement upon the addition of Br^-, Cl^-, F^-, SO_4^{2-}, and CO_3^{2-} in methanol (BTC = 1,3,5-benzenetricarboxylate). These anions weakened the vibrational quenching effect of OH group due to the interaction with hydrogen atom of methanol, thus enhances the luminescence intensity. In particular, F^- ion strongly interacted with the hydrogen atom, showing strongest significant enhancement of the luminescent intensity (Fig. 5).

Deng et al. demonstrated that transition metal ion-based MOFs can be used for the detection of anions [25]. The MOF of $[Cd(\mu_2\text{-}Cl)(\mu_4\text{-}5MT)]_n$ exhibits an emission at around 370 nm owing to the intra-ligand transitions (MT = 5-methyl-1*H*-tetrazole). The decrease in the emission intensity caused by the addition of NO_2^- is attributed to the energy transfer mechanism between the methyl groups and the nitrite ions.

4.3 Sensing of Small Molecules

The chemical functionality and structure of guest molecules significantly influence the luminescent emissions of MOFs. Rosseinsky et al. reported a pyrene-based MOF of $[In_2(OH)_2(TBAPy)]_n$ (TBAPy = 1,3,6,8-Tetrakis(benzoic acid)pyrene), which showed a wide range of solvent-dependent luminescence [26]. The emission band for the framework was shifted to lower energy by the desolvation. The emission intensity and the lifetime were dependent on the amount and the chemical nature of the species hosted in the pores. Polar DMF produced a high degree of structural order and separates the linkers to minimize self-quenching phenomena, resulting in a fine emission profile. Thanks to the hydrogen-bonding of the polar water guest and carboxylate groups, a framework structure was distorted, resulting in distinct emission properties (e.g., different properties are found using DMF as a solvent). In particular, the structural distortions induced by the interaction with polar guests strongly affected the lifetime of the emission. Different nonpolar molecules such as dioxane and toluene guests induced a limited distortion of the framework structure; however, as confirmed by diffraction analyses, the specific π-stacking interaction between the toluene and the pyrene moiety results in the reduced emission intensity.

The MOFs based on luminescent open metal sites exhibit attractive sensing capability for solvent molecules. Chen et al. reported the luminescent intensity change of an MOF induced solvent molecules coordinating an open metal site [27]. The MOF of $[Eu(BTC)(H_2O)_3]_n$ possesses an open metal site of Eu^{3+} (BTC = benzenetricarboxylate), which acts as a recognition site for guest molecules. While open Eu^{3+} metal site coordinated by DMF molecules showed an increased luminescent intensity related to the $^5D_0 \rightarrow {}^7F_2$ transition, the coordination operated by acetone to the metal site significantly decreased the intensity. Chen et al. demonstrated a further example of the luminescent intensity change triggered by the coordination of guest molecules to open metal sites [28]. An MOF based on $\{[Yb(BPT)(H_2O)] \cdot (DMF)_{1.5}(H_2O)_{1.25}\}_n$, (BPT = biphenyl-3,4′,5-tricarboxylate) shows an open metal site once guest molecules and coordinated water molecule are removed. The desolvated MOF of $[Yb(BPT)]_n$ exhibited emission from the $^2F_{5/2} \rightarrow {}^2F_{7/2}$ transition of the Yb_{3+} ion. As reported in the previous work [27], DMF and acetone most significantly enhanced and quenched the luminescence, respectively.

Hsu et al. demonstrated the sensing property of an $Na[Tb(OBA)_2]_n$ (OBA = 4,4′-oxybis(benzoate)) MOF [29]; the proposed MOF showed the ability to act as a stimuli responsive luminescent material towards a variety of solvents. The luminescence measurements of the MOF displayed that the $^5D_4 \rightarrow {}^7F_4$ emissions were the strongest intensities in BuOH and EtOH suspensions, but these intensities were significantly reduced in MeOH and H_2O suspensions. The small molecules of H_2O and MeOH get closer to the terbium ions of the framework that quenches the luminescence intensities effectively; on the other hand, the alkyl sterical hindrance of EtOH and BuOH potentially protects the terbium ions from quenching by OH oscillators.

In general, OH oscillators in the water molecules act as quenching species for the lanthanide luminescence. However, Wang and Gao et al. utilized water molecules to enhance the luminescent intensity of lanthanide-based MOFs [30]. The as-synthesized MOFs of $\{[Ln_2(fumarate)_2(oxalate)(H_2O)_4] \cdot 4H_2O\}_n$ (Ln = Eu, Tb) accommodated water molecules as guest species and showed luminescence derived from the lanthanide ions. The dehydration of the MOFs significantly reduced the crystallinity with a partial loss of the luminescent properties; after a rehydration process the luminescence was recovered. This is probably due to the softened environment of lanthanide ion in the dehydrated amorphous phases and the recovery of the stiffness after rehydration. A similar sensing behavior has been discovered by Li et al. using needle shape crystals of $\{[Eu_2L_3(H_2O)_4] \cdot 3DMF\}_n$, ($L = 2',5'$-bis(methoxymethyl)-[1,1′:4′,1″-terphenyl]-4,4″-dicarboxylate) [31]. The proposed Eu-based MOF presents an emission spectra corresponding to the $^5D_0 \rightarrow {}^7F_0$, $^5D_0 \rightarrow {}^7F_1$, $^5D_0 \rightarrow {}^7F_2$, and $^5D_0 \rightarrow {}^7F_4$ transitions. Remarkably, the MOF shows an eightfold enhancement of the luminescence intensity in presence of DMF vapors. The luminescent signal becomes weaker once the MOF is exposed to water vapors. Thus, the authors demonstrated sensor response of the MOF to DMF and water vapors. The vapors of DMF and water were introduced into the sensor cell in an alternating manner, and the luminescent intensity was monitored continuously. The turn-on and turn-off switching of luminescence was achieved within 10–20 s.

Molecules containing nitro or thiol groups often serve as efficient quencher because of the electron withdrawing properties or the hole trapping ability, respectively. Ghosh et al. reported a luminescent MOF of $[Cd(NDC)_{0.5}(PCA)]_n$ (NDC = 2,6-napthalenedicarboxylate, PCA = 4-pyridinecaboxylate) for the highly selective detection of the nitro explosive TNP (TNP = 2,4,6-trinitrophenol) [32]. The proposed MOF exhibited selective detection of TNP regardless the presence of other nitro compounds in both aqueous and organic solutions. The selectivity is ascribed to electron and energy transfer mechanisms, as well as electrostatic interactions between TNP and the luminescent center. The order of the quenching efficiency was found to be TNP 2,4-DNP > NP, which is in agreement with the order of acidity of the molecules (2,4-DNP = 2,4-dinitrophenol, NP = 4-nitrophenol). TNP with the highly acidic hydroxy group interact strongly with the luminescent center and efficiently quench the luminescence due to the favorable electron and energy transfer mechanism. Falcaro et al. reported luminescent hybrid materials by the encapsulation of QD in MOF-5, which detect thiol molecules by the quenching effect [8, 33]. Only thiols with the molecular size smaller than the pore were observed to affect the luminescence of the QDs. These studies suggested that the encapsulation of luminophores in MOF is also a successful method to fabricate molecular sieve MOF-based sensors.

The structural flexibility of MOFs, which allows for the confinement of molecules with maximized framework-guest interaction, is a prominent feature that differentiates MOFs from other conventional porous solids. Although the MOF dynamic structure holds promise for applications in storage and separation, MOFs have shown the potential to be used as an efficient chemical sensing platform, thanks to the framework's ability to adapt the nanopore depending on the analyte features. Furukawa and Kitagawa et al. demonstrated the guest-induced luminescence of dpndi-based interpenetrated MOF of $[Zn_2(bdc)_2(dpndi)]_n$ (bdc = 1,4-benzendicarboxylate, dpndi = *N*,*N*′-di(4-pyridyl)-1,4,5,8-naphthalenetetracarboxydiimide) [34]. The interpenetrated framework undergoes the dynamic structural transformation to confine a class of aromatic volatile organic compounds. The displacement of two chemically non-interconnected frameworks provides space for trapping the guest molecules. The confinement of guest species enhances the interaction of aromatic guest molecules with the dpndi embedded in the framework

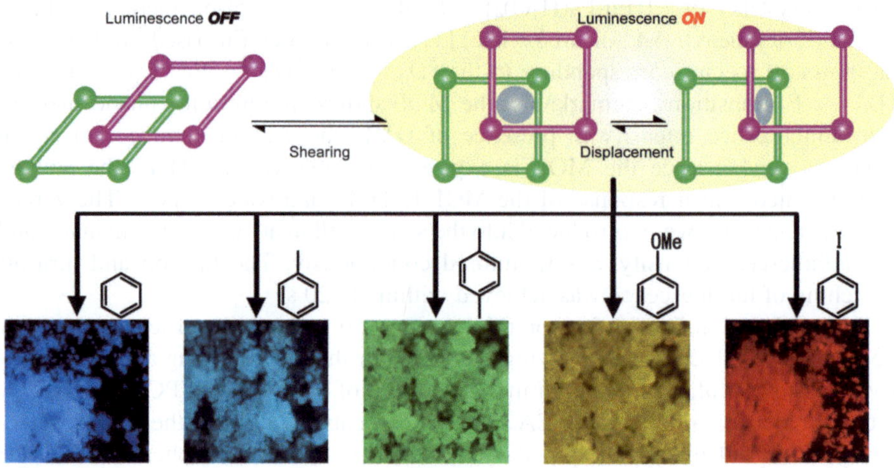

Fig. 6 The resulting luminescence of crystal powders, suspended in each liquid after excitation at 370 nm: benzene (*blue*), toluene (*cyan*), xylene (*green*), anisole (*yellow*), iodobenzene (*red*)

scaffolds. This leads to the formation of complexes with a CT nature, producing fluorescence from the exciplex or the stabilization of the triplet state of dpndi to generate phosphorescence. Although the MOF of $[Zn_2(bdc)_2(dpndi)]_n$ shows a very weak fluorescence with a very low quantum yield, the interaction between dpndi and aromatic guest molecules generated the visible fluorescence from violet to yellow by increasing the donor guest ability, as shown in Fig. 6. Furthermore, the interaction between dpndi and iodobenzene led to red emission by phosphorescence stabilized by heavy atom effect. Since the emission color depends on the chemical substitution on the phenyl ring, the MOF can identify the aromatic molecules by visible fluorescence.

Chen et al. established a new methodology to detect large biomolecules by using an MOF of $[Cu(H_2dtoa)]_n$ (H_2dtoa = dithiooxamide) [35]. In the absence of MOF, free probe DNA with a luminophore at its 3′ end has strong luminescence at an excitation wavelength of 480 nm. The introduction of MOF bound with probe DNA as a result of hydrophobic and π-stacking interactions between the nucleobases and the MOF quenched the fluorescence by an energy transfer process. In the presence of a target such as target DNA or protein the binding between the target and probe DNA alters the conformation of probe DNA. This structural alteration releases probe DNA from the MOF, which results in the restoration of fluorescence. Based on this sensing platform, HIV DNA and thrombin can be detected. The system has shown a high sensitivity, in fact the detection limits are 3 nM for the HIV DNA and 1.3 nM for the thrombin.

4.4 Sensing of Gas Molecules

There have been many examples of gas detection that uses luminescence quenching by energy transfer between gases and luminescent materials. However, only a limited number of gases, such as O_2 and NO_2, cause luminescence quenching. Thus, not so many studies on gas detection have been reported in luminescent MOFs. Lin et al. demonstrated O_2 induced luminescence quenching of $[Zn_4(\mu_4\text{-}O)(Ir(ppy)_3 \text{ derivatives})_2]_n$ [36]. The MOF exhibited the luminescence from $Ir(ppy)_3$ (ppy = 2-phenylpyridine) moieties. The MLCT phosphorescence can be readily quenched by molecules with a triplet ground state through the incorporation of O_2 into the framework. Thus, the luminescence of MOF is reversibly quenched by O_2. Petoud and Rosi et al. reported lanthanide doping in bio-MOF via cation exchange process [7]. The lanthanide-doped MOF also showed the capability of O_2 sensing.

Uemura and Kitagawa et al. demonstrated an MOF-based detection of CO_2 [37]. The photoactive molecules of distrylbenzene (DSB = distrylbenzene) were introduced into the MOF of $[Zn_2(bdc)_2(dabco)]_n$. The framework structure of MOF was distorted by accommodating DSB, and the structural conformation of DSB was twisted. Because the twisting of DSB affects the π-conjugation efficiency across the benzene rings, the MOF accommodating DSB did not show the luminescence. However, CO_2 adsorption into the extra space of pores induced the structural transformation of the MOF from distorted framework to original tetragonal framework. This structural transformation caused the conformational transformation of DSB from the twisted arrangement to the original linear one. As a result, the CO_2 adsorption triggers the luminescence emission of MOF-DSB hybrid materials. The flexible MOF and DSB synergistically collaborated to detect CO_2.

5 Conclusion

Luminescent MOFs are promising materials for chemical sensing, light-emitting devices, and biotechnology; however, the research on luminescent MOFs is still at an early stage. Although the host–guest-dependent luminescence mechanism related is now under exploration and new rational design of MOFs for specific sensing applications is becoming a raising research field, the majority of the literature reports are limited to the presentation of the luminescent properties. The designability of framework structures enables the manipulation of their inherent luminescent properties. Furthermore, the nanoporosity of MOFs allows for guest accommodation enabling the exploitation of guest-dependent luminescent properties. In the near future, if the luminescent sensing properties would be combined with the sieving function of the porous MOFs, sophisticated luminescent porous materials would be fabricated [38]. Moreover, the detailed mechanism of guest-dependent luminescence is going to be revealed thanks to the availability of fluorescence facilities with the ability to measure specific features including excitation spectra, lifetimes, or quenching rate constants. Moreover, the miniaturization of spectrometers make very promising the perspective to use portable sensing devices based on MOFs. The close multidisciplinary collaboration among chemists, materials scientists, photophysists, and engineers will certainly facilitate the implementation of some promising luminescent MOFs and technologies for practical applications in chemical sensing.

References

1. Allendorf MD, Bauer CA, Bhakta RK, Houk RJT (2009) Luminescent metal–organic frameworks. Chem Soc Rev 38:1330–1352
2. Cui Y, Yue Y, Qian G, Chen B (2012) Luminescent functional metal–organic frameworks. Chem Rev 112:1126–1162
3. Falcaro P, Furukawa S (2012) Doping light emitters into metal–organic frameworks. Angew Chem Int Ed 51:8431–8433
4. Fang QR, Zhu GS, Jin Z, Ji YY, Ye JW, Xue M, Yang H, Wang Y, Qiu SL (2007) Mesoporous metal–organic framework with rare etb topology for hydrogen storage and dye assembly. Angew Chem Int Ed 46:6638–6642
5. Müller M, Devaux A, Yang CH, De Cola L, Fischer RA (2010) Highly emissive metal–organic framework composites by host–guest chemistry. Photochem Photobiol Sci 9:846–853
6. Luo F, Batten SR (2010) Metal–organic framework (MOF): lanthanide(III)-doped approach for luminescence modulation and luminescent sensing. Dalton Trans 39:4485–4488
7. An J, Shade CM, Chengelis-Czegan DA, Petoud S, Rosi NL (2011) Zinc-adeninate metal–organic framework for aqueous encapsulation and sensitization of near-infrared and visible emitting lanthanide cations. J Am Chem Soc 133:1220–1223
8. Falcaro P, Hill AJ, Nairn KM, Jasieniak J, Mardel JI, Bastow TJ, Mayo SC, Gimona M, Gomez D, Whitfield HJ, Riccò R, Patelli A, Marmiroli B, Amenitsch H, Colson T, Villanova L, Buso D (2011) A new method to position and functionalize metal–organic framework crystals. Nat Commun 2:237
9. Wagner BD, McManus GJ, Moulton B, Zaworotko MJ (2002) Exciplex fluorescence of $\{[Zn(bipy)_{1.5}(NO_3)_2\}] \cdot CH_3OH \cdot 0.5pyrene\}_n$: a coordination polymer containing intercalated pyrene molecules (bipy = 4,4′-bipyridine). Chem Commun 2176–2177
10. Tanaka D, Horike S, Kitagawa S, Ohba M, Hasegawa M, Ozawa Y, Toriumi K (2007) Anthracene array-type porous coordination polymer with host–guest charge transfer interactions in excited states. Chem Commun 3142–3144
11. Zhang Z, Xiang S, Rao X, Zheng Q, Fronczek FR, Qian G, Chen B (2010) A rod packing microporous metal–organic framework with open metal sites for selective guest sorption and sensing of nitrobenzene. Chem Commun 46:7205–7207
12. Zhang X, Ballem MA, Ahrén M, Suska A, Bergman P, Uvdal K (2010) Nanoscale Ln(III)-carboxylate coordination polymers (Ln = Gd, Eu, Yb): temperature-controlled guest encapsulation and light harvesting. J Am Chem Soc 132:10391–10397
13. Zhang X, Ballem MA, Hu ZJ, Bergman P, Uvdal K (2011) Nanoscale light-harvesting metal–organic frameworks. Angew Chem Int Ed 50:5729–5733
14. Wu P, Brand L (1994) Resonance energy transfer: methods and applications. Anal Biochem 218:1–13
15. Xiao Y, Wang L, Cui Y, Chen B, Zapata F, Qian G (2009) Molecular sensing with lanthanide luminescence in a 3D porous metal–organic framework. J Alloys Comp 484:601–604
16. Bai Y, He GJ, Zhao YG, Duan CY, Dang DB, Meng QJ (2006) Porous material for absorption and luminescent detection of aromatic molecules in water. Chem Commun 1530–1532

17. Lu WG, Jiang L, Feng XL, Lu TB (2009) Three-dimensional lanthanide anionic metal–organic frameworks with tunable luminescent properties induced by cation exchange. Inorg Chem 48:6997–6999
18. Xiao Y, Cui Y, Zheng Q, Xiang S, Qian G, Chen B (2010) A microporous luminescent metal–organic framework for highly selective and sensitive sensing of Cu^{2+} in aqueous solution. Chem Commun 46:5503–5505
19. Liu W, Jiao T, Li Y, Liu Q, Tan M, Wang H, Wang L (2004) Lanthanide coordination polymers and their Ag^{+}-modulated fluorescence. J Am Chem Soc 126:2280–2281
20. Chen B, Wang L, Xiao Y, Fronczek FR, Xue M, Cui Y, Qian G (2009) A luminescent metal–organic framework with Lewis basic pyridyl sites for the sensing of metal ions. Angew Chem Int Ed 48:508–511
21. Zhao B, Chen XY, Cheng P, Liao DZ, Yan SP, Jiang ZH (2004) Coordination polymers containing 1D channels as selective luminescent probes. J Am Chem Soc 126:15394–15395
22. Zhao B, Chen XY, Chen Z, Shi W, Cheng P, Yan SP, Liao DZ (2009) A porous 3D heterometal–organic framework containing both lanthanide and high-spin Fe(II) ions. Chem Commun 3113–3115
23. Wong KL, Law GL, Yang YY, Wong WT (2006) A highly porous luminescent terbium-organic framework for reversible anion sensing. Adv Mater 18:1051–1054
24. Chen B, Wang L, Zapata F, Qian G, Lobkovsky EB (2008) A luminescent microporous metal–organic framework for the recognition and sensing of anions. J Am Chem Soc 130:6718–6719
25. Qiu Y, Deng H, Mou J, Yang S, Zeller M, Batten SR, Wue H, Lie J (2009) In situ tetrazole ligand synthesis leading to a microporous cadmium-organic framework for selective ion sensing. Chem Commun 5415–5417
26. Stylianou KC, Heck R, Chong SY, Bacsa J, Jones JTA, Khimyak YZ, Bradshaw D, Rosseinsky MJ (2010) A guest-responsive fluorescent 3D microporous metal–organic framework derived from a long-lifetime pyrene core. J Am Chem Soc 132:4119–4130
27. Chen B, Yang Y, Zapata F, Lin G, Qian G, Lobkovsky EB (2007) Luminescent open metal sites within a metal–organic framework for sensing small molecules. Adv Mater 19:1693–1696
28. Guo Z, Xu H, Su S, Cai J, Dang S, Xiang S, Qian G, Zhang H, O'Keeffe M, Chen B (2011) A robust near infrared luminescent ytterbium metal–organic framework for sensing of small molecules. Chem Commun 47:5551–5553
29. Lin YW, Jian BR, Huang SC, Huang CH, Hsu KF (2010) Synthesis and characterization of three ytterbium coordination polymers featuring various cationic species and a luminescence study of a terbium analogue with open channels. Inorg Chem 49:2316–2324
30. Zhu WH, Wang AM, Gao S (2007) Two 3D porous lanthanide–fumarate–oxalate frameworks exhibiting framework dynamics and luminescent change upon reversible de- and rehydration. Inorg Chem 46:1337–1342
31. Li Y, Zhang S, Song D (2013) A luminescent metal–organic framework as a turn-on sensor for DMF vapor. Angew Chem Int Ed 52:710–713
32. Nagarkar SS, Joarder B, Chaudhari AK, Mukherjee S, Ghosh SK (2013) Highly selective detection of nitro explosives by a luminescent metal–organic framework. Angew Chem Int Ed 52:2881–2885
33. Buso D, Jasieniak J, Lay MDH, Schiavuta P, Scopece P, Laird J, Amenitsch H, Hill AJ, Falcaro P (2012) Highly luminescent metal–organic frameworks through quantum dot doping. Small 8:80–88
34. Takashima Y, Martínez VM, Furukawa S, Kondo M, Shimomura S, Uehara H, Nakahama M, Sugimoto K, Kitagawa S (2011) Molecular decoding using luminescence from an entangled porous framework. Nat Commun 2:168
35. Zhu X, Zheng H, Wei X, Lin Z, Guo L, Qiu B, Chen G (2013) Metal–organic framework (MOF): a novel sensing platform for biomolecules. Chem Commun 49:1276–1278
36. Xie Z, Ma L, deKrafft KE, Jin A, Lin W (2010) Porous phosphorescent coordination polymers for oxygen sensing. J Am Chem Soc 132:922–923

37. Yanai N, Kitayama K, Hijikata Y, Sato H, Matsuda R, Kubota Y, Takata M, Mizuno M, Uemura T, Kitagawa S (2011) Gas detection by structural variations of fluorescent guest molecules in a flexible porous coordination polymer. Nat Mater 10:787–793
38. Hirai K, Furukawa S, Kondo M, Meilikhov M, Sakata Y, Sakata O, Kitagawa S (2012) Targeted functionalisation of a hierarchically-structured porous coordination polymer crystal enhances its entire function. Chem Commun 48:6472–6474

Index

MIX
Papier aus verantwortungsvollen Quellen
Paper from responsible sources
FSC® C105338

If you have any concerns about our products,
you can contact us on
ProductSafety@springernature.com

In case Publisher is established outside the EU,
the EU authorized representative is:
Springer Nature Customer Service Center GmbH
Europaplatz 3, 69115 Heidelberg, Germany

Printed by Libri Plureos GmbH
in Hamburg, Germany